Deepwater Oil Production and Manned Underwater Structures

MICHAEL E. JONES

Marine Technology Programme
The University of Glasgow

Graham & Trotman

First published in 1981 by
Graham & Trotman Ltd.
Bond Street House
14 Clifford Street
London W1X 1RD

British Library Cataloguing in Publication Data

Jones, Michael E.W.
Deep water oil production and manned underwater structures.
1. Gas well drilling, Submarine
2. Oil well drilling, Submarine
I. Title
622'.338 TN880.2

ISBN 0-86010-339-0

ISBN 0 86010 339 0

Printed in Great Britain by J.W. Arrowsmith Ltd, Bristol
Typeset in Great Britain by Academic Typing Service,
Gerrards Cross, Bucks.

CONTENTS

LIST OF FIGURES

LIST OF TABLES

INTRODUCTION

This book is primarily an investigation into the use of large manned one atmosphere underwater structures for the production of oil and gas from the deepwater areas of the world. However, to establish a firm base for the detailed study it was considered necessary to review the prospects for such resources in deepwater areas and to determine the limitations of present and proposed offshore production systems. This book consists of four major sections, A,B,C,D, which are sub-divided into chapters relevant to the subject matter.

Section A reviews the prospects for oil and gas resources in deepwater in a worldwide context and attempts to identify prospective areas and potential operational water depths. Technological aspects of present offshore production operations and future capabilities are identified. The potential applications of present, proposed and future production systems are analysed in relation to water depth and the case for further investigation of the manned subsea concept made.

A technical assessment of the problems of placing men and machinery on the seabed in a one atmosphere environment is undertaken in Section B. A multi-disciplinary approach is used to assimilate knowledge from past habitat designs and related technology i.e. nuclear submarines, space etc. Life support systems, submersibles, power supplies, physiology, safety etc. are covered in some depth. The technical feasibility of the concept is then evaluated from this work.

Section C provides detailed systems analysis of the topics covered in Section B. Operating systems and parameters of interest are displayed diagrammatically. Interactions between various subsystems in the context of the total system are analysed diagrammatically, and some relevant factors indicated. Other technology that will have a significant impact on the concept is discussed and other work being undertaken on one atmosphere structures is considered. Various scenarios are then generated to suggest the possible direction in which future subsea production configurations may develop.

In general, a summary or overview is given at the end of each section or chapter as indicated in the Contents. Section D provides general conclusions from the work, recommendations for future research and development requirements and an overall summary.

SECTION A

Prospects for Oil and Gas in Deepwater and Offshore Production Technology

1

PROSPECTS FOR OIL AND GAS IN DEEPWATER

Large commercial volumes of oil are rare and their occurrence is difficult to predict. This is especially true of deepwater areas where the geology is not well known, although the Deep Sea Drilling Project is providing useful information. However, consideration can be given to general information regarding the deep water provinces in an attempt to make an assessment of the likelihood of the occurrence of commercial hydrocarbon deposits.

Oil and gas are found in sedimentary basins, the prospectivity of a discovery being determined by the quantity of hydrocarbons generated in the sediments and the quality and size of the reservoir rocks. The quality of the rock must be of sufficient thickness, porosity and permeability and large enough to justify commercial exploitation. The physical generation of oil and gas is determined by an adequate supply of organic carbon which has been heated to a sufficiently high temperature by burial to achieve its generation threshold. Oxygen deficiency or reducing conditions are required for oil generation.

The thickness or depth of the sediment will determine whether oil or gas is formed. Oil generation usually occurs in a sediment depth range of 1 km to 3.5 km, with an optimum for giant fields being 1.5 km to 3.5 km. Gas is abundant at depths of 1 km to 4.5 km and on down to 10 km.[2] Oil generation occurs at depths giving a temperature range of 70–110°C where oil generation reaches a maximum. No oil is found below the 'oil floor', where gaseous hydrocarbons are produced as a result of oil 'cracking'. The zone between the threshold for generation and oil floor is called 'mature' sediment and below this 'over-mature' sediment. Deep sea drilling cores will show richness of hydrocarbon source rocks and the degree of maturity (Fig. 1.1).[1]

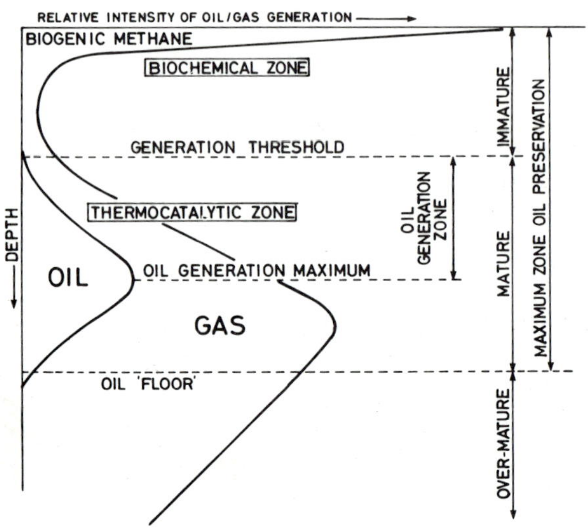

Figure 1.1 Generalized scheme of hydrocarbon generation in sediments

Investigation of sedimentary deposits in the world oceanic area reveals that the Mid-Ocean ridges, which account for 32.7% of the area have virtually no sedimentary cover and can be discounted of any hydrocarbon potential. The Abyssal Plains or Ocean Basins, which cover 41.8% of ocean floor are generally featureless plains and at depths greater than 4000 m, the cover of sediments is too thin for hydrocarbon formation, although coral reefs may give some gas production. Three quarters of the world's oceanic area therefore shows no real promise of hydrocarbon potential.[2]

The Continental Margins (Fig. 1.2) do exhibit the necessary thick sediments along the edges of the major land masses from which they are derived. They include the main deep water areas that may have significant prospects of producing hydrocarbons in commercial quantities. The Continental Margin is considered as the 'slope' from the edge of the Continental Shelf Break at approximately 200 m out to the edge of the land derived sediment, synonymous with continental 'rise'. Oil and gas deposits in the shallow areas of the continental shelf can be traced out into the deeper waters.[3]

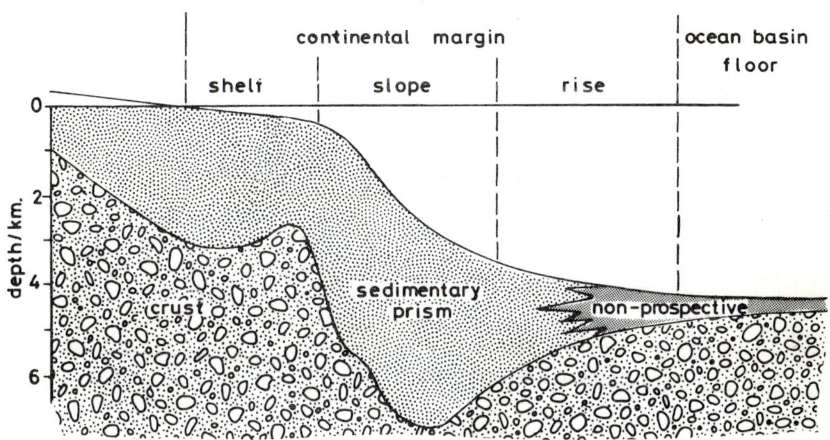

Figure 1.2 Schematic cross section through a passive continental margin.

Oil sedimentary basins occur geologically in three characteristic forms, Convergent, Divergent and those remote from continental margins at their time of formation, the Interior Continental Plate Basins. Offshore we are concerned with Convergent and Divergent basins, specifically passive or pull-apart types (Divergent) in Atlantic, Indian and Arctic Oceans and collision margins (Convergent) off California. The California Offshore structures are rather unique, so the divergent basins would appear to hold the best prospects (Fig. 1.3).

Considering the shallow water sediments of the Atlantic Slopes, as there is no oil or gas ashore, there are no grounds to expect any different offshore, present deposits in the North Sea being in a unique tectonic environment. The deep water sediments of the Atlantic Slope could, however, be an important domain for hydrocarbon prospects by the nature in which the sediments have been laid down. There seems to be good reason for confidence that

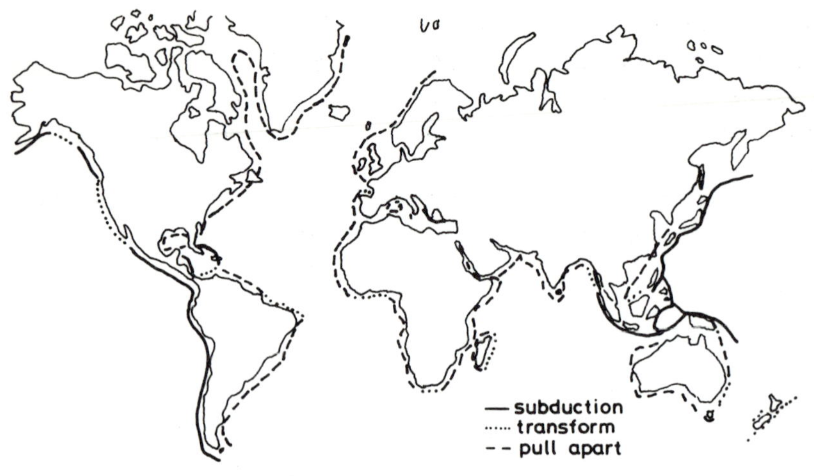

Figure 1.3 Continental margin types.

these deepwater sediments will contain oil and gas, some of which will be large enough to justify the cost of deepwater exploration and production (Fig. 1.4).[2]

In a worldwide context, the margins of the Indian Ocean would also appear to hold similar prospects to the Atlantic continental margins. Australian margins are also promising, especially the North West Shelf which offers large structures, which have good potential for hydrocarbons, although there is more likelihood of gas finds. The Arctic Ocean also has potential but involves the complication of the harsh environment. The Pacific Ocean margins have narrow shelves and the descent to deep waters is rapid and although sediments of the required thickness occur they have poor reservoir characteristics and economic fields will be hard to find and unlikely to exceed those of the Atlantic and Indian Oceans. The enormous sediments at the mouth of larger rivers, i.e. Niger, Indus, offer potential if reservoir conditions are right and Niger basin exploration has been encouraging.

Of today's world oil, divergent basins produce 23%, convergent basins 9% and plate interior basins 68%. The plate interior basins (Middle East etc.) are relatively accessible while most divergent and convergent basins are offshore and in harsh environments. Oil production distribution today would appear to be a function of the ease of exploration and development of different types of margin. Figure 1.5 shows that the majority of the world's prospective areas are non-productive to date and also mostly unexplored.[4] The majority of the undrilled basins exist in harsh explorative climates of the Arctic, deep oceans, jungles and areas of extreme logistic support difficulties. It indicates that there is as much oil to be found as has been found to date and that much of it could come from the harsh land environment as well as the offshore continental margins. The tendency, however, will be towards

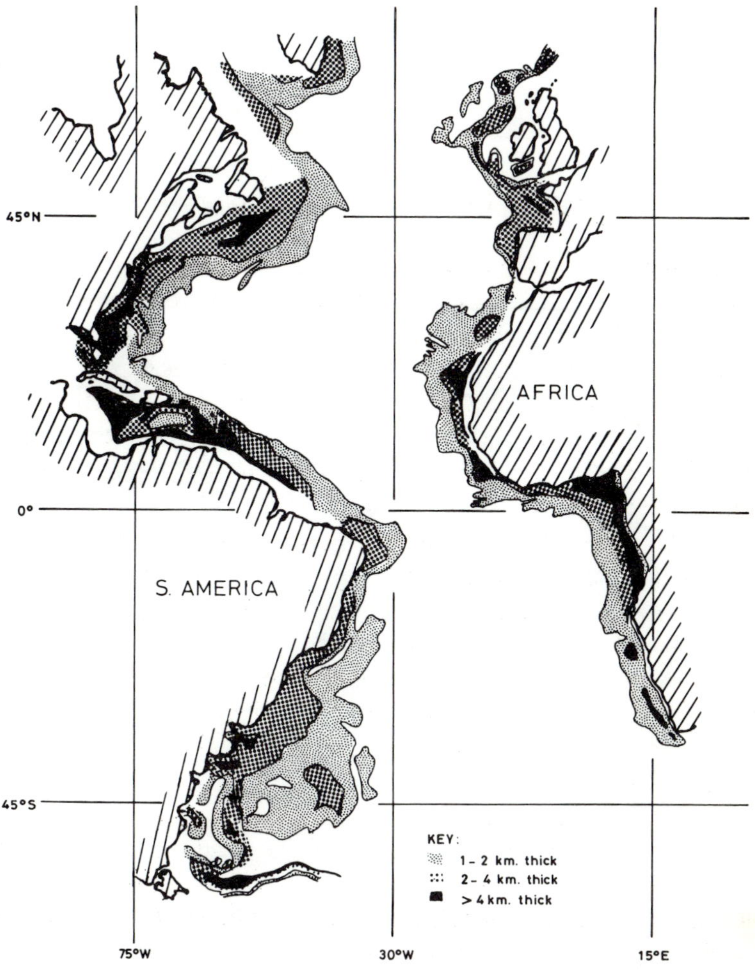

Figure 1.4 Depth of sedimentary deposits in the Atlantic Ocean.

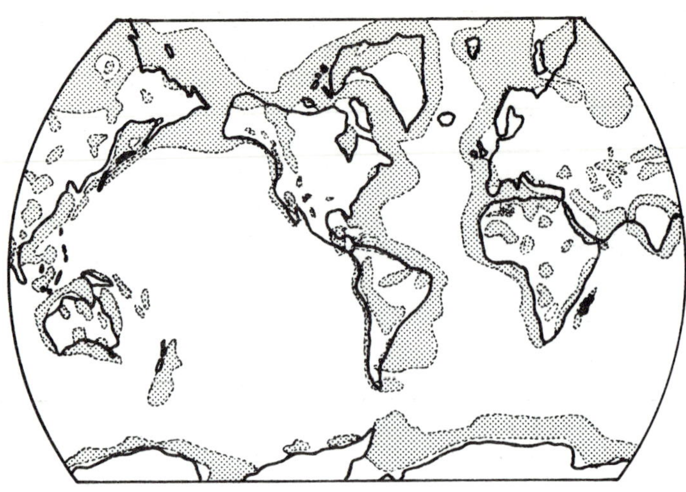

Figure 1.5(a) World prospective areas—frontier and semi-frontier basins.

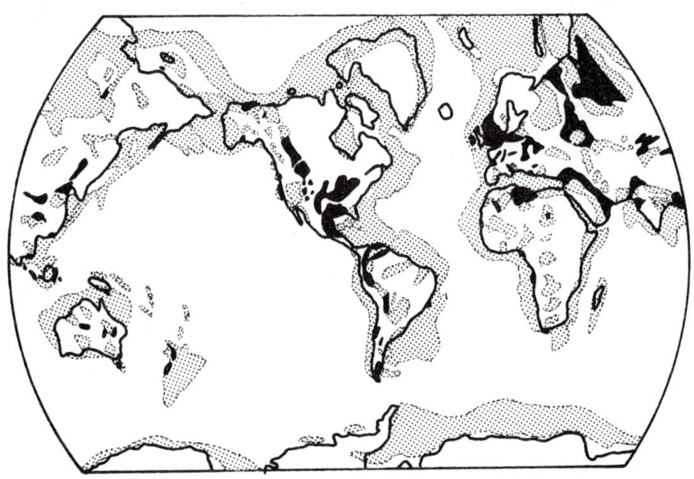

Figure 1.5(b) World prospective areas — ▒ sedimentary basins and ■ productive basins.

larger numbers of smaller fields rather than a smaller number of giants.

Exploration drilling is at present being carried out in the following offshore areas (see also Fig. 1.6):

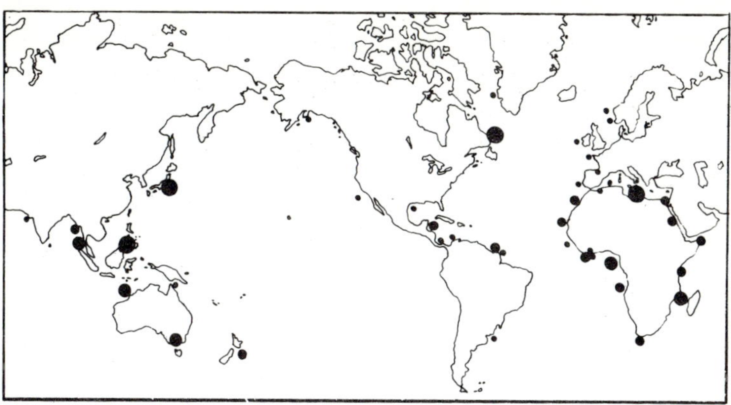

Figure 1.6 Deep water drilling activity at more than 200m. (1977).

1. Offshore Eastern Canada and Newfoundland.
2. Offshore the Eastern seaboard of the United States/The Baltimore Canyon.
3. The Beaufort Sea, Offshore Alaska and North West Canada.
4. The Yellow Sea, and the South China Sea, Offshore Shanghai and Hong Kong.
5. Offshore North Western Australia and the Exmouth Plateau off Western Australia.
6. Offshore Argentina, near Tierra del Fuego.
7. Offshore West Africa.
8. The Atlantic, West of Ireland, North and West of Shetland and off Norway North of 62 parallel.
9. The Gulf of Mexico, off the United States and the Bay of Campeche off Mexico.
10. The continental shelf off Venezuela and off Brazil.
11. The Western Mediterranean.

Recent discoveries include oil in 80 m of water two hundred miles southeast of St. Johns, Newfoundland, on the Grand Banks[7], gas in the Baltimore Canyon and deepwater discoveries in the North Sea and Celtic Sea. Shell have discovered a very large gas accumulation in 323 m water depth off Norway and British Petroleum an oil well in 450 m water depth West of Eire which is rumoured to be a big structure of potential significance.[8] These two discoveries mean the realisation of a new frontier in development technology as the industry is confronted with exploitable fields in water depths beyond the reach of conventional field development technology. A test well was drilled in 1980 in the North Rockall Trough northwest of the Hebrides, (BNOC) which may lead to developments in 2000 m water depths in the next decade.[6,9]

In summary, the large areas of the deep ocean basins have little prospect of containing large hydrocarbon reserves. The Continental margins down the rise clearly have prospects but the incidence of large fields that will warrant production will be limited. There is certainly enough promise to encourage exploration for deep water sources and significant oil finds will undoubtedly be made in deep water; it will be difficult and expensive to find and extract. Many prospective land areas of the world also remain unexploited, especially in the hostile environments. The potential oil base is very large, but exploration and development in deep waters will be critical for oil supplies into the next century. The next twenty years will probably see the exploitation of offshore resources to depths approaching 2000 m on the continental slope. Production facilities will need to be developed to meet this requirement.

2

INTRODUCTION TO TECHNOLOGICAL ASPECTS OF DEEPWATER PRODUCTION

The search for oil is moving into progresssively deeper water; all the major oil companies have extensive deepwater exploration licences. The lead time of new oil discoveries until production is of the order of ten years. Some estimates indicate that fifty per cent of new oil discoveries will come from offshore fields. The deeper and more hostile environment will demand advanced technology for exploration and production with the inherent high costs and risks of such ventures.

The financial risks offshore are high and are determined by the environment (water depth, location and seabed characteristics), reservoir quality (reserves, transmittability, shape, pay depth and crude quality) and the proposed transportation system for the product. All eventual requirements of the field development have to be included in the original project design for the most economic development and this indicates the importance of a full field appraisal before commitment to a development scheme. Offshore developments are characterized by a major commitment at the beginning without an opportunity for pilot testing of reservoir conditions. Production of oil/gas is required as early as possible to alleviate the financial burdens of the initial investments for such developments.

The search for oil in the deeper waters of the U.K. will be on the shelf, slope and rise of the Continental Margin and in depths of water that will range from 300 m to 2000 m (north of 62°N,

South Western Approaches, Faeroes Trough and Rockall Trough and Plateau).[6]

If major resources are discovered in the large areas of the north western waters of the U.K., which run off the shelf into very deep water, the exploitation of these resources will require a completely new and costly technology.[5] Worldwide prospects indicate that similar technology will be required for producing hydrocarbons from other deepwater areas of the world (Chapter 1).

The technology that will be developed in the future will be directed by the economic and social considerations that determine the feasibility of developing deepwater resources. The necessary marine technology will be very expensive to develop, but due to the international nature of the offshore industry there should be adequate resources. There will be distinct needs for better technology, but the cost of the equipment will need to pass the economic test at the price of oil at the time. Technology will probably advance by staged development, rather than by a technological breakthrough because of the high financial risks involved. The introduction of exotic unproven technology could cause a financial breakdown, if the new systems were unsuccessful.

A brief review is given of present offshore technology in relation to exploration and production systems. The aim is to assess the capabilities and limitations of present systems and the prospects for the application of proposed concepts for deepwater exploitation.

2.1 DRILLING

Rotary drilling methods are almost universally used today with mud lubrication and turbo drills for directional drilling. Drilling in shallow waters is done from bottom standing platforms and in deep waters from semi-submersibles or dynamically positioned drillships. Today's fully dynamically positioned drillships are capable of operating in 2000 m water depths and can drill 8000 m below the rotary table. Sonar and closed circuit television systems are used to give wireless re-entry to the wells.

The offshore depth record to date (May 1979) is held by Offshore International's Discoverer Seven Seas' 1486 m well at a location 205 nautical miles east-northeast of St Johns, Newfoundland.[14] Five of this vessel's last eight wildcats have been driven in water depths greater than 1000 m. Esso's Sedco 472 drilling programme off the northern coast of South America in water depths of 1280 m has involved the added complication of operating in currents of up to 3 knots. Glomar Challenger, while investigating geological phenomena, drilled in water depths of 6140 m. Well depths have increased to the order of 3000 m. Glomar Explorer is at present being converted for the Ocean Margin Drilling Program (U.S.) and will be capable of drilling on the continent margins in water depths as great as 7620 m. In 4000 m water depth it will be able to drill 7000 m below the seabed.[18]

There are many problems involved in deepwater drilling. Well control devices placed subsea have to operate with perfect reliability and the dynamic motion of the vessel or semi-submersible has to be allowed for at the top of the drill string. Facilities must be available to allow quick disconnection of the drill string in an emergency or in bad weather conditions. Strong currents impose additional problems, due to the high stresses imposed on the riser near the surface where the current is strongest producing bending moments. Currents also complicate the re-entry operations into the well, as the riser will be deflected by current drag forces. Any deepwater drilling location has its own special operational requirements, flexibility and proper technical pre-planning will in most cases meet the circumstances.[15]

More emphasis is now being placed on equipment that can handle higher working pressures while drilling offshore. In an attempt to cut costs improved drilling techniques and methods of monitoring of downhole conditions are being developed (two exploration wells in the North Sea cost £10million, i.e. £50 per inch). There have been several recent advances in the field, i.e. B.O.P. (blow-out preventer) stacks that are rated to 15,000–20,000 p.s.i., downhole safety valves, guidelineless drilling allowing quick disconnect and other developments that are improving offshore drilling techniques.[16,17]

The capability to drill efficiently and safely over the upper continental slope and in relatively strong currents exists. Dynamic positioning of vessels on location for extended time periods is established. B.O.P. technology and auxiliary support systems for the most part have been proven to be well designed and reliable. The riser-drillstring/vessel interface is the most critical area. Special deepwater drilling technology that is adaptable to specific operating conditions is evolving and should not be a restriction to the development of deepwater hydrocarbon deposits in a hostile offshore environment.

2.2 RISERS

The riser is the vital production link between the surface and subsea systems. It transmits well products upwards to the surface and oil export and gas/water injection downwards and routes work-over operations and control lines to activate remote subsea systems. Risers were initially a direct development of drilling riser technology, but now flexible hose alternatives are available. The riser system is the weak link between the sophisticated subsea and surface hardware. The choice of a suitable marine riser is a major problem area, complicated by the prospect of deeper and more exposed waters.

The segmented vertical tension riser is an extension of drilling experience. The central steel tube carries the oil export and has tension applied near to the top of the riser to support its weight and to increase the pipe's resistance to lateral deflection and buckling. Around the central tube are the flowlines for each well and a bundle of control lines to activate the subsea systems. This arrangement is termed a dispersed riser system. In an integral riser system all the lines are contained in a protective tube to reduce inherent problems with deep water applications.[32]

The design of the riser is affected by many factors; the riser's natural buoyancy, tension force to be applied, currents and depth at location, wave loading, particularly at the splash zone, and platform motion. The riser connection to the subsea and surface hardware can be of several alternative designs, but at no time must the riser act as an additional anchor for the production system. The operating philosophy will determine the type of attachments fitted. If the riser system is designed to survive the worst storm condition, production will be shut down but the risers left permanently attached. If this is not the case attachments will need to be disengaged in adverse conditions and the risers, flowline and control lines retrieved to the deck of the platform.

The achievement of a reliable riser system for deepwater work would appear to require improvements in the following areas:

1. High integrity mooring system to restrict the movement of the platform structure to within riser tolerances in varying sea states and water depths.
2. Permissible stress in the riser is often determined by the fatigue life. Smaller, more frequent waves cause most fatigue so the riser design should have suitable response at these frequencies.
3. High pressure flowlines and long-life flexible hoses.
4. Reduction in 'clashing' of risers due to tension differences and hydrodynamic interactions, by the use of bundles etc.
5. Methods of riser retrieval, handling and storage.
6. Vortex shedding characteristics.
7. Related work on umbilicals and methods of diverless subsea connections.

The development of riser technology for use with floating production systems is at an early stage. The permanent and reliable operation of such systems appears to require much improvement for operation in water depths to 500 m. In deeper water, unless integral buoyancy is given to the riser enormous tension loads will be required to support the riser weight. Induced dynamic loads and vortex shedding will impose additional forces on the riser. The logistics of riser handling, umbilicals and diverless subsea connections will involve another dimension of complication. The simple extension of present designs for deepwater application would appear to be in doubt.

2.3 SUBSEA SYSTEMS

The use of subsea completion systems is now a well established concept with proven operational experience. Earlier systems tested in the less hostile conditions of the Gulf of Mexico and elsewhere are now being applied to the deeper waters and more exacting environment of the North Sea and Arctic. The technique is expensive, at present the cost of drilling, completing and 'hook-up' of one underwater completion can be up to ten times as expensive as the cost of a conventional platform well. The industry, however, recognizes it as a means of ensuring optimum reservoir drainage under North Sea conditions. Developments are proceeding rapidly in such critical areas as control equipment and tie-in methods.[5]

Subsea systems are finding application in many diverse roles, depending on the configuration and reserves of the field under development. They can be used for early production on any size field, for initial production and testing of the reservoir conditions of marginal fields and as satellite wells feeding to fixed structures from outlying deposits or isolated accumulations too small to justify individual platforms. In proposals for floating production systems for the exploitation of deepwater and marginal fields subsea systems are an essential element of the concept. The wells may also be used for injection purposes if reservoir stimulation is required.[32]

2.3.1 Completions

Subsea completions developed to date are of three main types:

1. Wet and at ambient sea pressure — Ambient Diver Accessible
2. Wet and at 1 Atmosphere — Diver Accessible (Vickers Intertek-Neutrabaric)
3. Dry and at 1 Atmosphere — Accessible (CanOcean Resources Limited)

Operating and maintenance of the systems is a function of the environmental design. The wet ambient sea pressure system relies on remote control from the surface with the possibility of diver intervention if the operating depth is not too great. The new wet 1-atmosphere system involves diver transition to an enclosed wellhead capsule by submersible, and then transfer to a 1-atmosphere (water environment) in which work is undertaken on the wellhead by the diver.[25] The dry/1-atmosphere system involves intervention using a service capsule into an enclosed 1-atmosphere (air) wellhead capsule, and work is done in a shirt-sleeve environment.[27] The system used is only determined by personal preference, although the oil industry at present seems to favour wet completions (ambient pressure). This situation may change with the move to deeper waters.

The completed wells can be in clusters directly below a floating platform, individual satellite wells or clusters of satellite wells. The number of wells directly below the production facility is usually limited to 12 - 15 and sometimes, to get optimum drainage it is advantageous to distribute a smaller number of wells at strategic points, this limits the number of wells that have to be closed in during work-over. Satellite wells are usually fed to a manifold, rather than directly to the surface, to reduce the complexity of flowlines and to help with pressure losses due to long flowlines.

The back pressure on the well production system, especially with the use of satellite wells consists of losses in the well tubing, determined by the length and size of the lines. Losses occur in the tubing, i.e. vertical flowstring, in the horizontal flowstring and in the riser or vertical flowline. The lift to the surface depends on water depth and as this restraint increases with deeper water there is a need to consider first stage separation on the seabed.[3]

2.3.2 Templates

A cluster of deviated wells can be drilled through a template from a semi-submersible platform, the template needs to be accurately levelled to meet the required drilling tolerance and also fixed by piles. The template must also act as a base for production and export riser and be able to cope with vertical and horizontal movements of the floating production system on the surface and any flowline pull-in operations. It also supports pipework, valves etc. which must be vertically retrievable for servicing or replacement. In a more complex form it may also incorporate a manifold and control hardware.

2.3.3 Manifold

The manifold acts as a submarine flow station and brings together the oil flows from a number of wells, combines them and feeds them in a bulk line to the surface avoiding the use of many flowlines and risers. The design can be 'wet' or enclosed in a one atmosphere work space for 'shirtsleeve' repair or pull-in operations.[26] The manifold centre can stand alone as a collection point or may also incorporate a template for deviated drilling. Manifolds under development for the North Sea are very large (48m x 42m) to enable them to cope with the large daily flowrates of the fields.

2.3.4 Through Flowline Maintenance of Wells (T.F.L.)

This is a method that makes it feasible to reach the wells from the platform by passing tools through the flowline using a hydraulic circuit, by selecting the proper tool subsurface valves in the well-bore can be operated or exchanged.

The operation is monitored from the surface and the system can enter the well from a remote point, unlike the conventional method of running tools vertically down a wireline for which a semi-submersible is required. For North Sea flowrates at least 4 inch or larger T.F.L. tools will be needed.

Improvement in subsea system designs is aimed at increasing the reliability of subsea control devices and devising configurations that offer a minimum profile to collision with trawl-boards and other underwater dangers. The move is also towards automatic techniques to reduce dependence on diver intervention, especially with use in deeper waters, (> 300 m). The basic concept of subsea systems is well established. More attention is now being given to the safety of personnel involved in the operations and the safety of the operating environment.[23]

Subsea systems will proliferate in existing production areas to depths of 300 m for tapping the outlying parts of individual fields and adjacent minor reserves. The developments during this period will form the technological base for future subsea activities in deeper waters. The exploitation of deepwater deposits will be highly dependent on the successful development and operation of such systems.[22,28] (See also Section 5.15.)

3

INTRODUCTION TO PRODUCTION SYSTEMS

The physical environment and potential climatical extremes at the location of an offshore oil well will determine which type and strength of structure is required to carry the production facility. The oil quality, volume and method of extraction will establish what weight and volume of equipment will need to be supported and contained by the structure. The choice of a suitable structure will therefore be determined by the satisfaction of a complex set of requirements.

Oil, gas and/or water can be produced from a well under its own pressure, or the pressure may need to be artificially maintained by gas or water injection. A gas well will mainly produce methane with small amounts of ethane, propane, butane and acid gases. An oil well will produce both gas and heavier hydrocarbons. The separation of gas from an oil producing well can involve several stages of separation. The characteristics of each well will therefore determine individually the requirement for facilities, although some equipment will be common to all systems. The methods of storage and transport of the product will involve other equipment combinations.

The wells may be predrilled by a dedicated drilling rig or the structure may be required to act as a combined drilling and production structure. The structure could be a permanent production facility for a large or marginal field, or a temporary production system for early production of a large field.

The selection and design of an offshore production system will be determined fundamentally by the environment conditions, the characteristics of the well and financial restraints on the production time scale. The development method will be modulated by the experience of tried technology and a reluctance to use unproven technology because of the high financial risks involved.

The complex nature and worldwide variation in the conditions of oil exploitation would seem to indicate that any of the proposed systems in the following review could have potential application for oil production. The nature of the industry makes it inherently difficult to specify discrete areas of application for each system.

3.1 PRESENT PRODUCTION SYSTEMS

3.1.1 Fixed/Bottom Mounted Structures

The first generation platforms were steel-piled, single-purpose platforms that sat on the sea-bed. These were then developed into the multi-purpose steel platforms and concrete gravity platforms that are now in use. The majority of offshore oil is at present produced by the first and second generation fixed platforms. Recent improvements in offshore heavy lifting techniques and piling methods have been responsible for the current swing back to piled steel platforms combined with lift on modules in the U.K. Sector of the North Sea.[5] (See Fig. 3.1.1 a,b,c.) A gravity steel structure has also been proposed.

The overall feeling in the oil industry is that fixed structures have established a foothold in shallow water and are likely to retain application in this field. They are suitable for water depths up to 200 m in North Sea conditions and up to depths of 300 m in more sympathetic environments, such as the Gulf of Mexico. At greater depths costs and logistics are restrictive and there is the possibility of the flexibility and massiveness of such very tall structures being subject to resonance at frequencies of wave excitation. However, new concepts of fixed platforms, i.e. Condeep T300, MAN 400 etc. may extend fixed platform application in the North Sea to 300 m water depth (Fig. 3.1.1 d and e).[31]

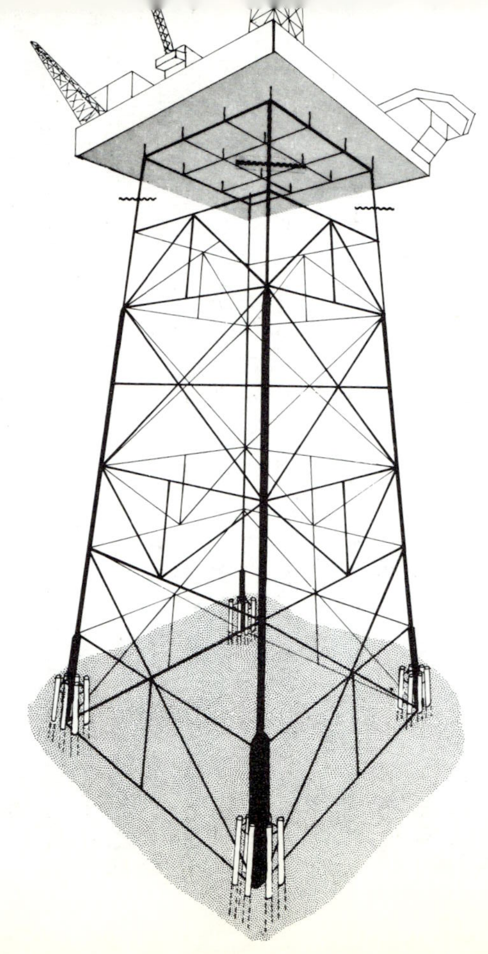

Figure 3.1.1(a) Fixed steel platform.

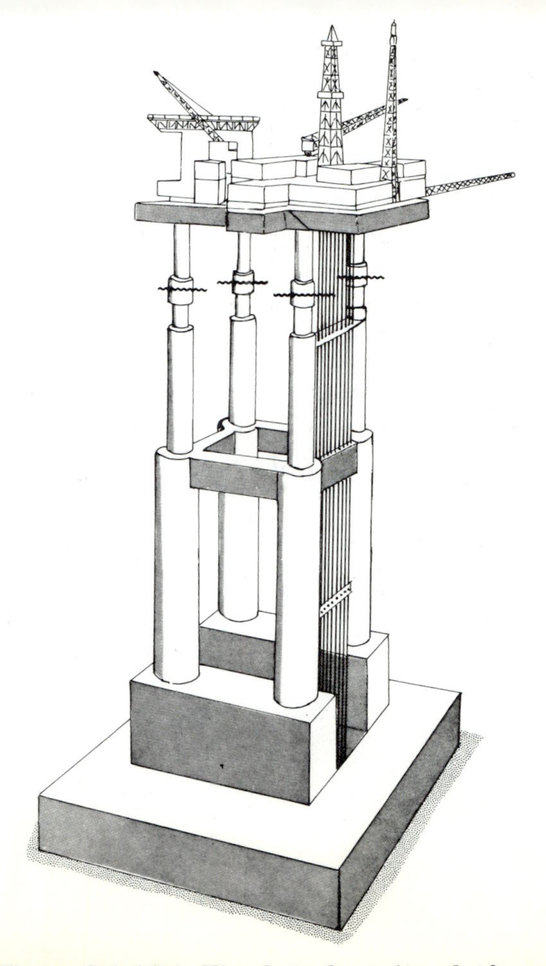

Figure 3.1.1(b) Fixed steel gravity platform.

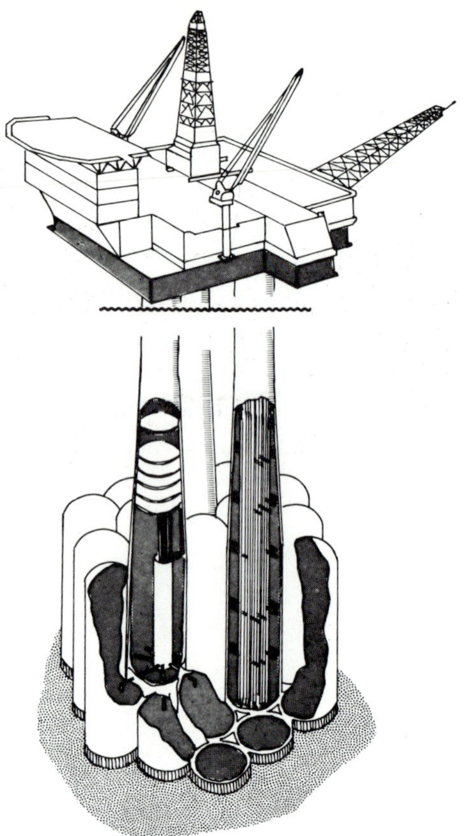

Figure 3.1.1(c) Shell/Esso Brent B platform — Condeep design.

3.1.2 Floating Production Systems

The floating production system can be considered the third generation of offshore production facility as it represents a fundamental change in offshore philosophy from fixed to floating systems. The basic concept can be considered as a buoyant platform structure moored over subsea wellheads with the two interfaced by a production riser (Fig. 3.1.2a).[33] The platform at present would be a converted tanker or semi-submersible, but future systems will be purpose designed for this application. These will be discussed in the next section.

The concept offers the potential of early production, deepwater

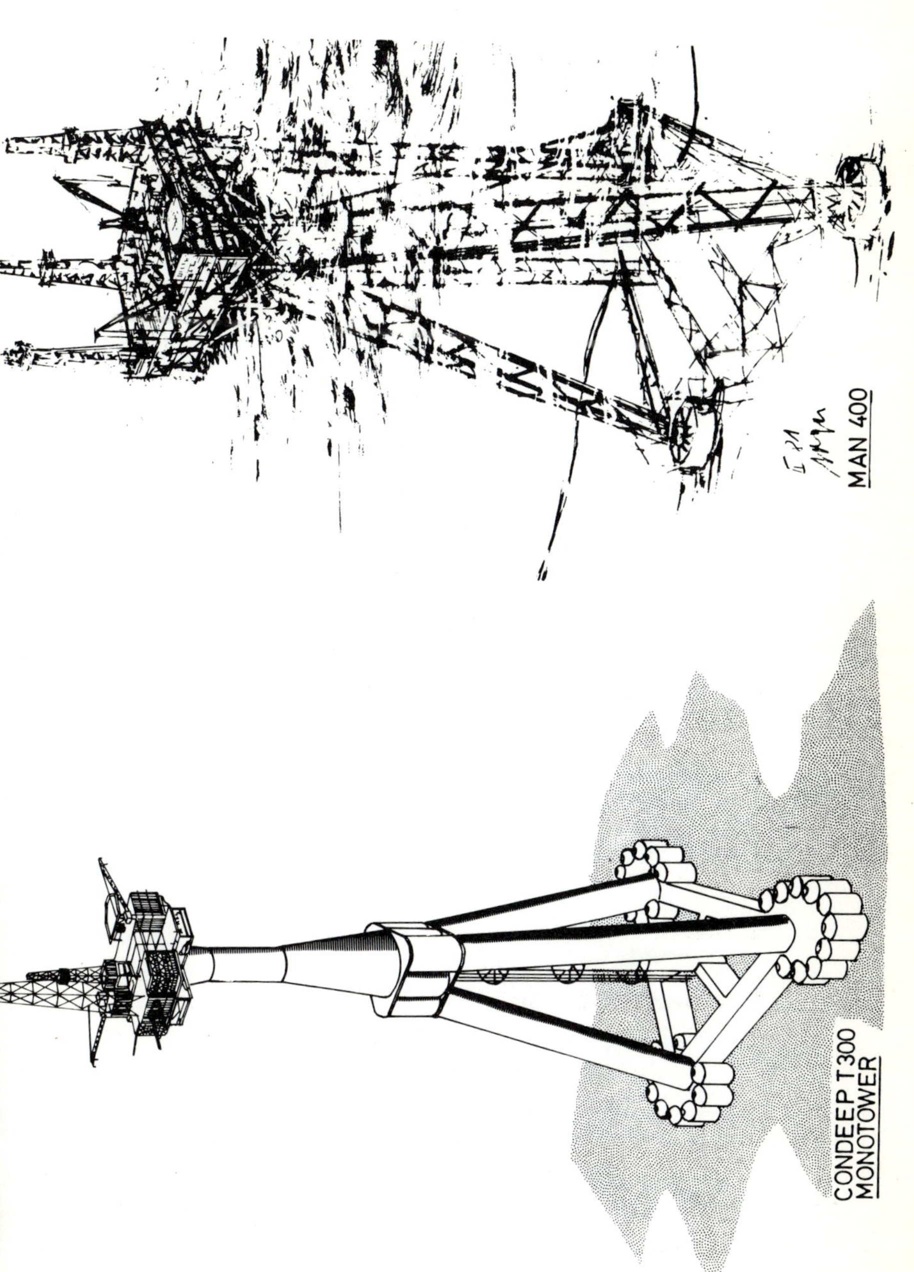

Figure 3.1.1(d) Condeep T300 monotower

Figure 3.1.1(e) MAN 400.

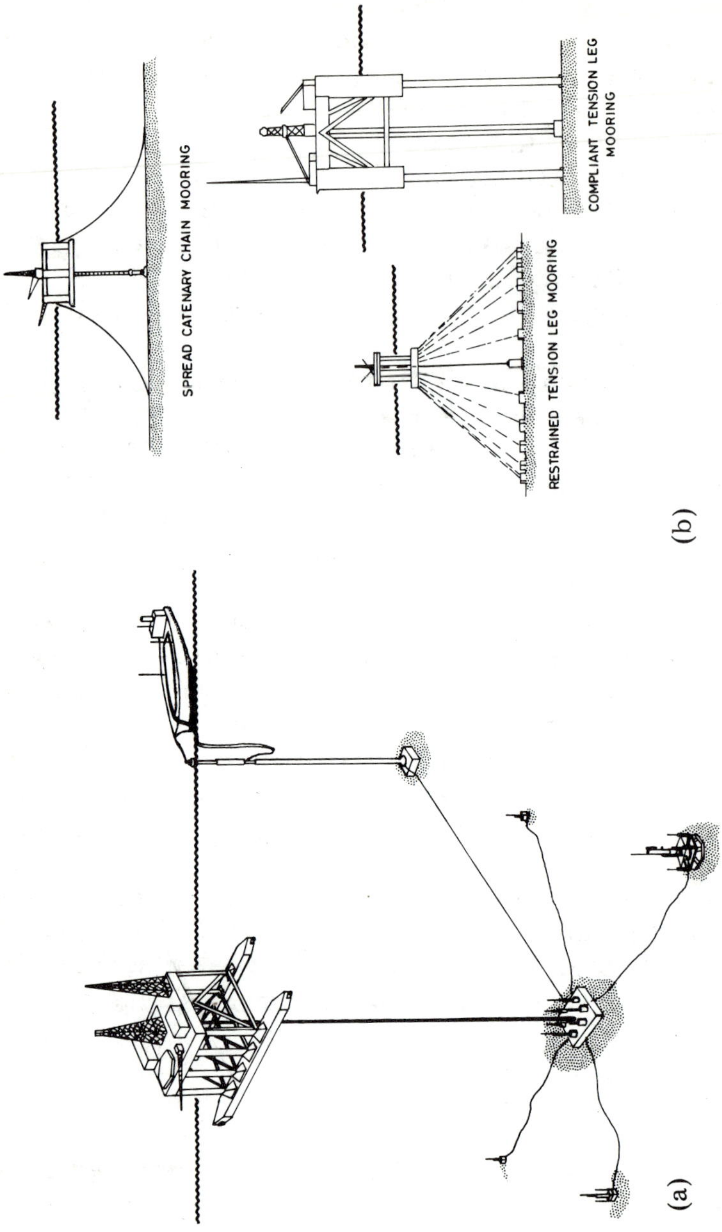

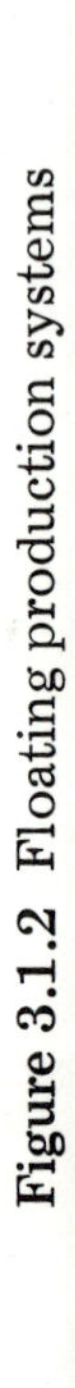

Figure 3.1.2 Floating production systems

applications and the development of marginal fields. On large fields the system would be used for early production and on marginal fields as a total production facility in association with subsea systems. The system has a minimal sensitivity to water depth and is relatively mobile allowing it to be moved to a new location after finishing its initial function. Oil export is usually by tanker loading from a buoy connected to a subsea export line. The limiting criterion for production is the restriction of tanker loading operations due to weather downtime, although buffer storage can be incorporated in the buoy system to alleviate this problem.

The first floating production system was installed in the North Sea in June 1975 by Hamilton Brothers on the Argyll field, using a converted semi-submersible, Transworld 58. The operational experience gained with this system over several years has established the floating production system as a viable field development method. Tanker loading has been the primary cause of production shut down.

The critical components of the concept are the riser connections and mooring system. Riser developments have been based on drilling riser techniques, but further work is being done on high-pressure, long-life flexible hoses. The riser must be designed to handle the environmental conditions it will encounter and must be easily disconnected in case of an emergency. The integrity of the riser depends on the capability of the mooring system to reduce movements of the platform. Soft mooring systems (spread anchor catenarys) are used and stiff mooring systems (compliant-tension leg) are under development (Fig. 3.1.2b). Anchoring of these systems in difficult soil conditions is another complication. The hydrodynamic performance of the floating production platform will be sensitive to the *loading* of the platform, and this needs to be controlled closely.[32]

The floating production system in association with subsea systems is now a tested concept, improvements are still required in riser, mooring and tanker loading systems. They may be used as a total production system for marginal fields or as an early production system for large fields. The Argyll field is situated in 100 m water depth and is probably the shallow limit for such a system although it is possible to moor a semi-submersible in 70 - 80 m of water providing the environmental conditions are favourable.

A maximum operating depth is difficult to predict because of present limitations of riser and mooring systems.

3.2 PROPOSED PRODUCTION SYSTEMS

The systems discussed are presently undergoing detailed design and testing, but are not necessarily in commercial operation. Many of the concepts are more sophisticated examples of the floating production systems discussed in the previous section. Most incorporate subsea systems and various combinations of sub-systems are used depending on the specific choice of anchoring system, riser design and transportation system. The operation and maintenance of the subsea system will also be a major characteristic of the system design.

3.2.1 The Guyed Towed (Exxon)

Exxon Production Research Company have carried out a three-year, large-scale, offshore test on a one-fifth scale model of the tower installed in 90 m water depth. These tests have formed the basis for the prototype design of a tower for operation in 350 m water depth in the North Sea environment. The guyed tower concept may extend the water depth limitation of bottom founded structures to at least 600 m and also permits conventional drilling and production operations from the deck structure. Crude oil is transported by a pipeline, or if in deepwater an offshore terminal may be used.[34]

The tower structure has a square cross-section which is constant over its length. One end pierces the surface and carries the deck equipment, while the seabed termination simply rests on a vertical bearing foundation, called a spud can (Fig. 3.2.1). The well conductors penetrate through the spud can. An array of mooring wires hold the tower upright and these run through swivelling fairleads to clump weights on the ocean floor and then extend on to piled or conventional anchors. The tower is 'compliant' and moves slightly in response to applied wave forces, the mooring arrangements providing sufficient restoring force to the tower so that it will essentially remain vertical under design conditions.

Since the tower pivots at the base, tower motion will induce

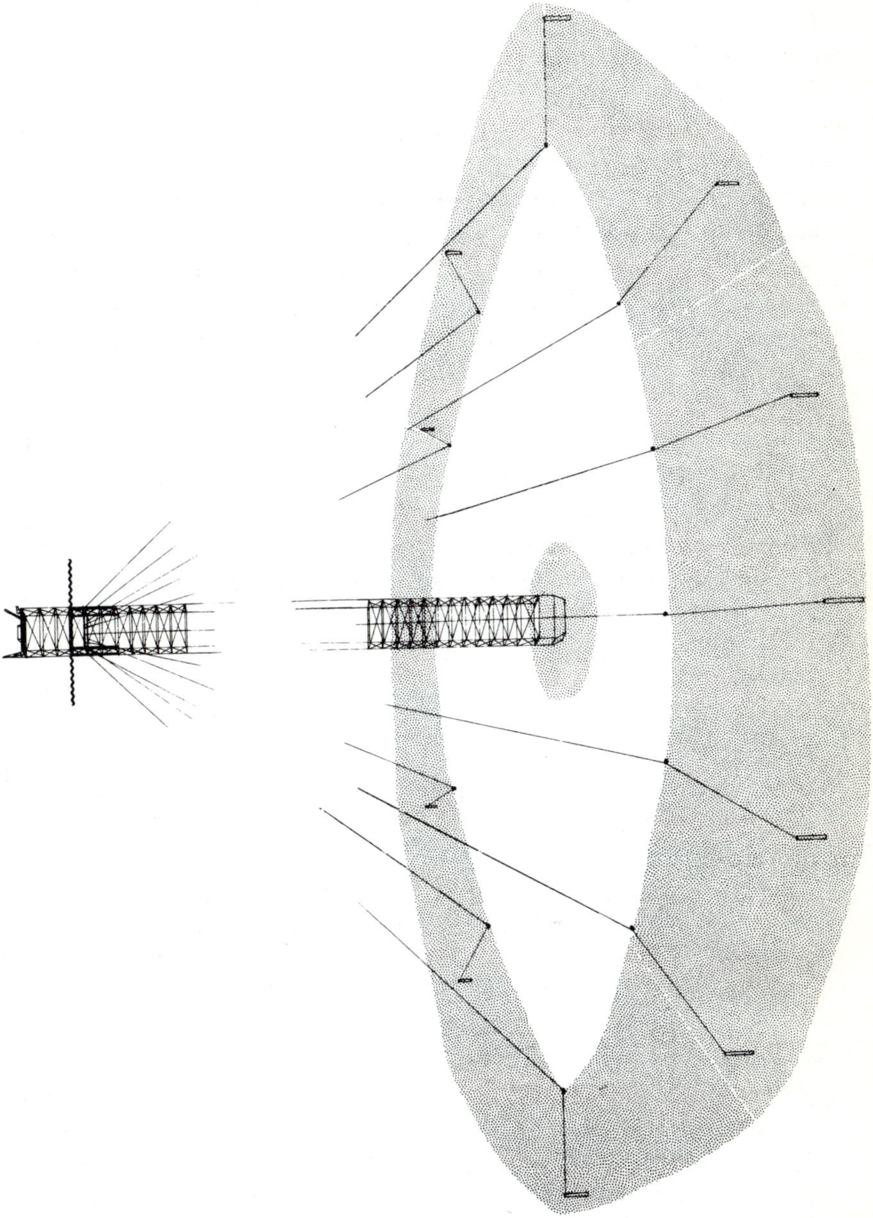

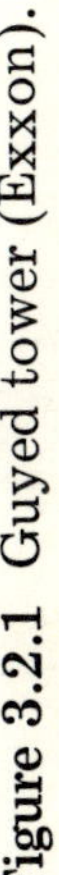

Figure 3.2.1 Guyed tower (Exxon).

flexure in the well conductors at the spud can. Studies have shown that conventional conductor designs are adequate if tilt is limited to an angle of about 2°. To retain this limit of angular tolerance, in less than about 200 m is impractical with the present mooring design.

The prototype design for the North Sea consists of a tower of cross-section of 36.5 m^2 and utilizes sixteen legs. The guideline system has 24 wire ropes of 4¼ inches, clump weights each of 165 tons and 1.4 m piles and would reach to a radius of approximately 1000 m at 350 m water depths. The system would accommodate 30 well conductors. The design would be limited to an operating depth range of 250 – 520 m by the constraints of the 2° tilt angle and six second flexural period, respectively. A new design could extend the depth limitation by using larger structural cross-sections for deeper waters and modifying mooring systems for shallower water operations.

Exxon have recently confirmed that it will make the first full-scale field test of its guyed tower concept in 1000 ft water depth at the Lena Field located 25 miles south of the Mississippi River mouth and 65 miles southeast of Grand Isle, Louisiana, near the site of the prototype guyed tower tests. It will probably be used for drilling and production and installed during the summer of 1983.[35] In this application a pile foundation is to be used instead of the spud-can foundation. The piles will be more than 1500 ft long, descending from 15 ft above water to the mudline and another 500 ft into the seafloor. The piles will be welded together as they are driven, and will act like a long spring allowing the tower to bend.[50]

3.2.2 Tethered Buoyant Platform (T.B.P.) (or Tension Leg Platform, T.L.P.) British Petroleum Limited, Vickers Offshore and Department of Energy. EEC

The system consists of a subsea wellhead system and a semi-submersible type floating structure moored with vertical tethers to the seabed. A riser is used for oil flow between the seabed and the T.B.P. and for servicing the well and reservoir. Tanker loading, or a pipeline system may be used to transport the export oil from the field. Considerable prototype design work has been carried out since 1975 on the system components and the next stage will involve development engineering for a specific scheme (Fig. 3.2.2).

The system was designed as a whole, taking into account the interaction between different sub-systems. Two fundamental

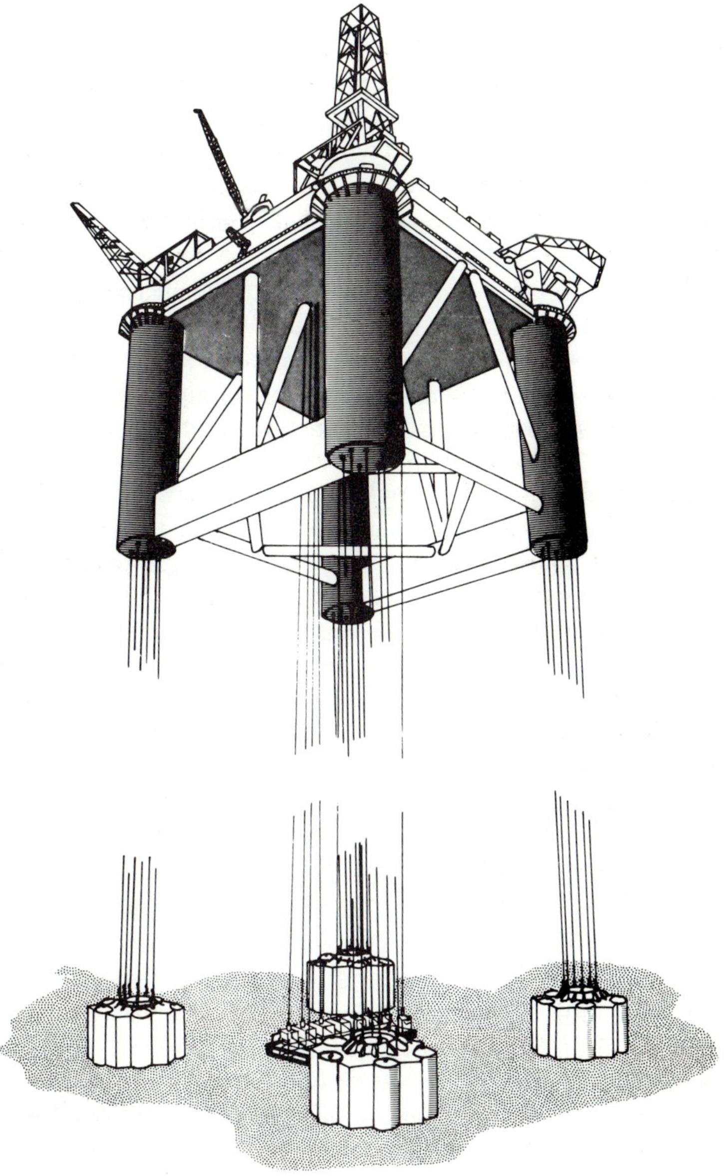

Figure 3.2.2 Tethered buoyant platform (British Petroleum).

requirements were assumed. One, that to save deck load the T.B.P. would not be used to drill production wells and the other that simple proven engineering would be used wherever possible.[36] The tether system restrains the platform in heave, but may exhibit significant surge and sway motions under wave and current action. Conventional catenary moored systems have restrained surge and sway but exhibit significant heave. Riser design has to be optimized for this movement in surge, which is a function of water depth.

The tethers consist of six separate high-tensile wires in each corner of the platform. The tethers are pretensioned at the time of installation by adjusting the length of the tethers and the draught of the T.B.P. The tension must be such that even under the worst sea-state they do not go slack, otherwise the wires will snap. The tethers run through the platform legs and connect to subsea anchors at each corner. The tension and condition of the cables is monitored and considerable redundancy is built into the tether system. The seabed anchors can be gravity or piled depending on soil conditions.

The riser is a tensioned vertical steel riser similar to those developed and proven for drilling systems. The tension on the riser is applied at the top to support the riser's weight and increase the pipe's resistance to lateral deflection. Various methods of attachment of the riser at the platform and subsea are possible. At shallow depths it is preferable to use separate risers but deeper (> 450 m) bundles and subsea manifolding may be advantageous.

The tension leg platform is a significant development of existing floating production platforms. The heave difficulties are reduced at the expense of sway, on the consideration that this movement is more easily dealt with by riser design. At depths approaching 100 m sway is significant so T.B.P.s are unsuitable for use in shallow waters. Investigations have shown that no problems are insoluble for water depths up to 450 m even for the most severe environment, but beyond this depth there may be problems of dynamic instability to be overcome.

The technical state of development of the system shows considerable promise as a production facility in the 100 – 450 m range in hostile environments, with the possibility of extension to 600 m water depth. The system was recently considered for the Magnus field development in 183 m of water; it was technically feasible but was not the most economic solution because of the large amount of separation facilities required, so a tall steel platform will be used.

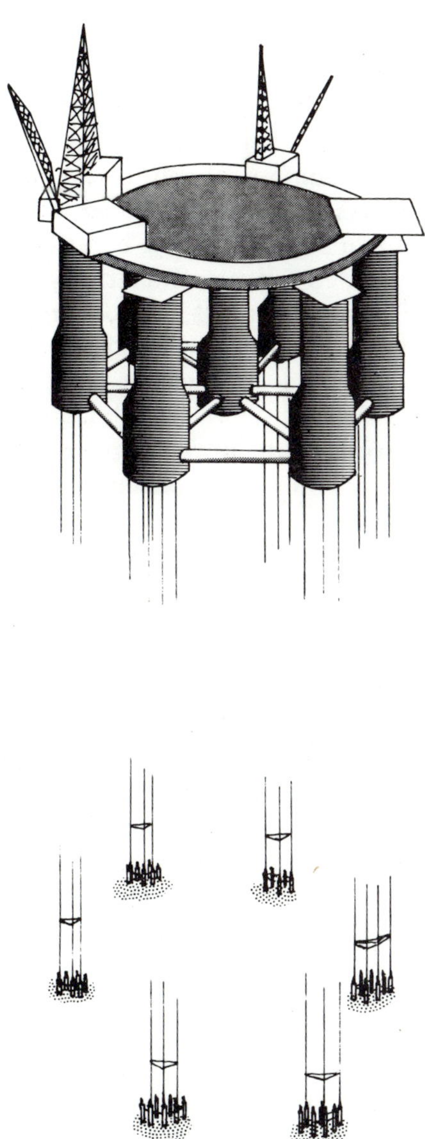

Figure 3.2.3 Tethered buoyant platform (Gulf Oil).

3.2.3 Tension Leg Platform (T.L.P.) Gulf Oil

A recent design of tension leg platform has been proposed by Gulf Oil in association with Earl and Wright and Seaflow Incorporated. A detailed feasibility study has been undertaken and technical development of the concept is proceeding. The system is considered suitable for application in water depths from 400 to 1000 m, depending on the location of the field.[37,38]

The platform structure consists of six peripheral buoyant columns and centre column. Each peripheral column houses four tension members that are anchored to seabed templates, through each template eight wells can be drilled. Catenary mooring lines are also provided for the installation phase (Fig. 3.2.3).

Production will be combined at a seabed manifold and brought to the surface through a single riser system at each template. Crude oil will be exported via a single point mooring system to a tanker.

Gulf is now investigating a rectangular configuration using either a subsea or deck completion system.

3.2.4 Tension Leg Platform (T.L.P.) Fluor Subsea Services/Scott Lithgow, EEC, Department of Energy, Several Oil Companies

The Deep Oil Technology prototype DPX-1 had a similar configuration to the T.L.P. system, but was characterized by three tether points rather than four. The one-third size prototype has been tested off Catalina Island, California, at depths to simulate conditions for commercial platforms in water depths of 210 m. These tests were completed in the summer of 1979.

Designs for service in the North Sea considered by Scott-Lithgow and their associates Fluor Subsea Services are based on a six-legged square T.L.P. structure (Fig. 3.2.4).

3.2.5 Vertically Moored Platform (V.M.P.) Amoco

The vertically moored platform is basically a semi-submersible type floating platform that can support drilling and production operations in deepwater.[40] The jacket includes four large buoyancy chambers which supports its own weight and the thirty-two riser pipes. The risers are suspended from the bottom of these

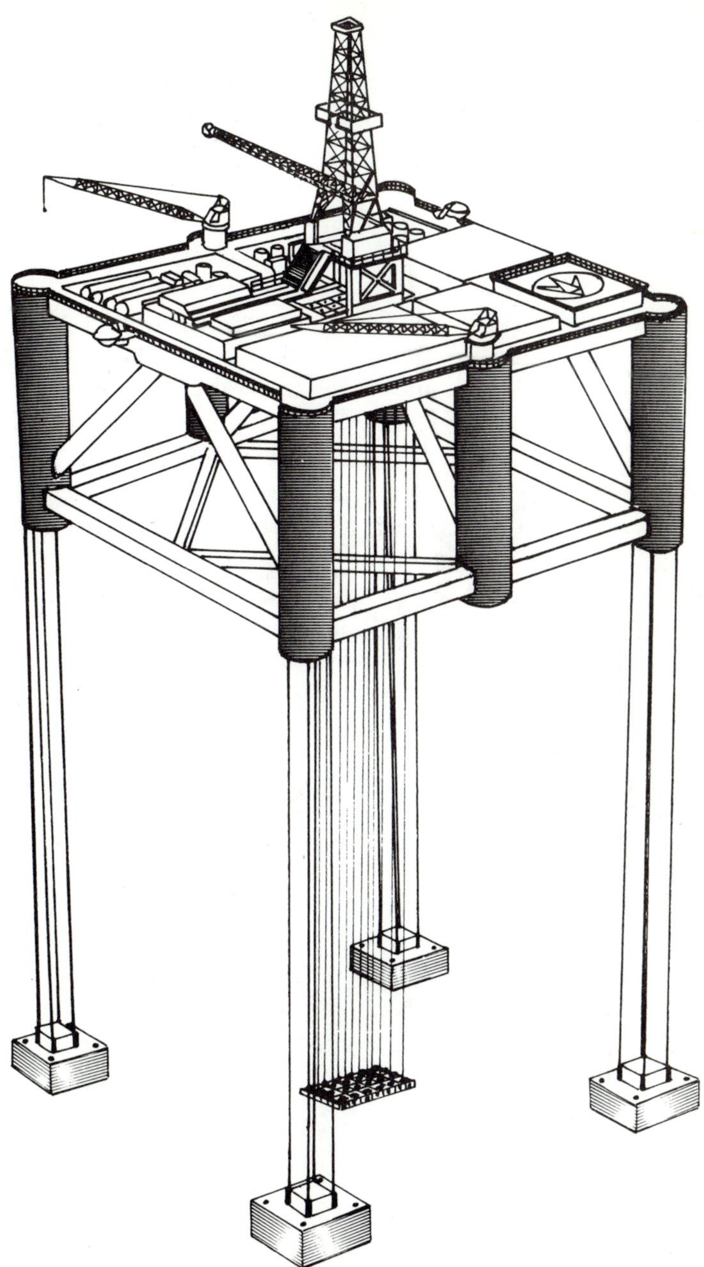

Figure 3.2.4 Tension leg platform (Fluor Subsea Services/ Scott Lithgow).

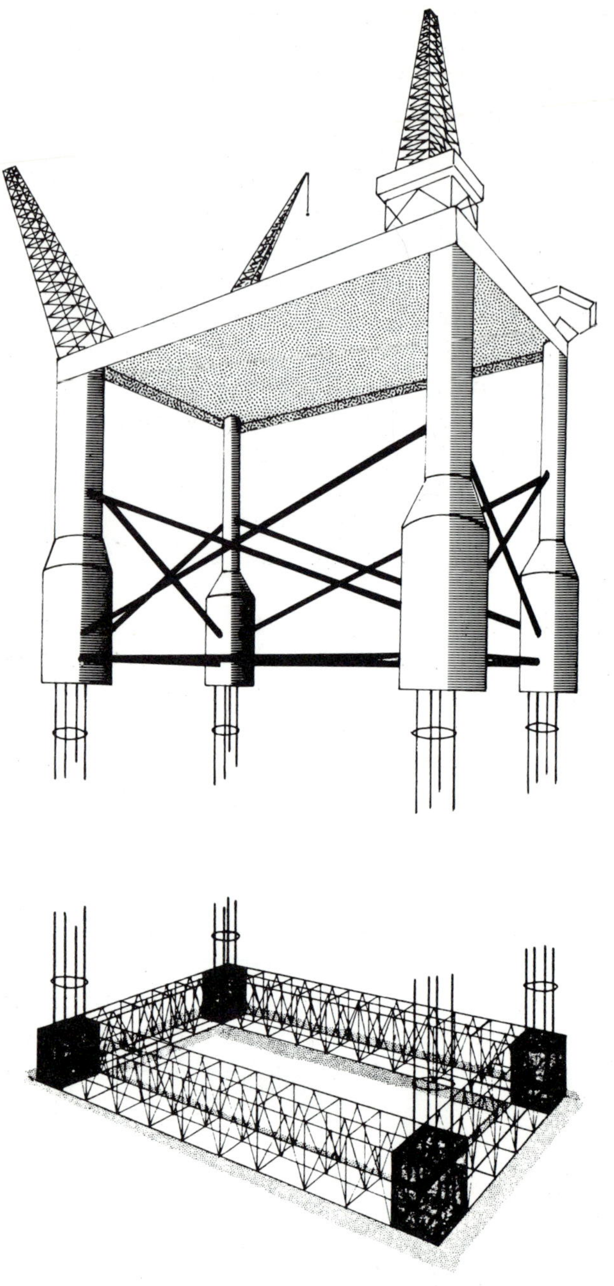

Figure 3.2.5 Vertically moored platform (Amoco).

chambers and extend to the seabed where they are cemented in place, excess buoyancy keeps the risers in tension. The riser system allows the platform to surge but effectively eliminates heave, pitch and roll (Fig. 3.2.5).

The integral structural riser system is the unique and key element of the V.M.P. concept. The structural riser itself is composed of seamless tubing and connectors for make-up of the individual sections. At the seabed the structural risers themselves are connected to the drive pipe and conductor system, which is cemented into the seafloor formations, the drive pipe and conductor acting essentially as a grouted tension pile. The risers are spaced with special frames to stop riser fluctuations and assist installation. The riser system allows the wells to be drilled and produced through the structural risers, with the Blow Out Preventer on the structure, and after completion the Christmas trees are located on deck. Well drilling and production operations from the V.M.P. are essentially the same as from a conventional fixed structure.

Detailed design studies and model tests have been carried out over a twelve year period. The results indicate that the system will have feasible and practical application in water depths from 250 m to 1000 m. The lower depth limit is heavily influenced by the environment, less hostile environments would probably allow use to 160 m. The economic operating range would appear to be 330 - 1000 m.

3.2.6 Tension Leg Platform (T.L.P.) Conoco

The tension leg platform will be a massive form of semi-submersible and will use sixteen vertical mooring tethers of about 10 inch diameter pipe with 3½ inch thick walls to connect it to the foundations on the seafloor. Each of the tethers will be made up like a string of drill-pipes with specially torqued joints. The foundations will be fastened to the seabed by piles that are driven in and cemented. The wells will be drilled through the well template (Fig. 3.2.6).[41,42]

The mooring system will be maintained in tension by ballasting the platform. Special flex joints, similar to ball joints but made of a special sandwich of rubber and steel will be used at each end

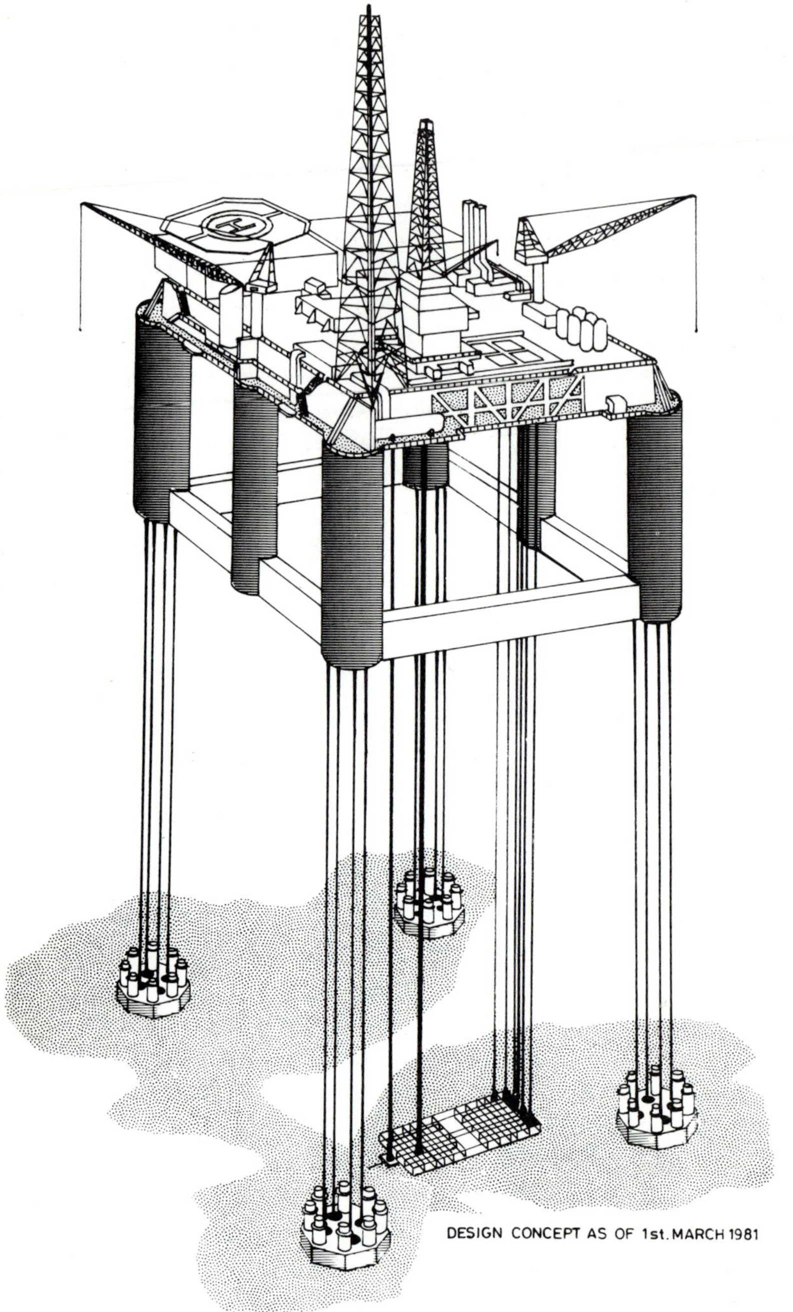

Figure 3.2.6 Tension leg platform (Conoco).

of the mooring tethers to connect with the platform and foundations. The flex joints allow 20° of travel. An innovative locking device at the end of the sixteen mooring tethers makes it possible to seat and lock the tether at the seafloor by stabbing it into the anchor assembly. The latch is permanent, but can be released for inspection purposes or removal of the T.L.P. The risers are maintained in tension by passive tension jacks.

The design concept is claimed to have application in water depths far in excess of that possible with existing platforms. The Conoco system will be the first commercial application of the tension leg concept in the world. It is proposed to use the design in the Hutton Field in the North Sea in 147 m water depth. The depth is toward the shallow limit for the application of such a concept, however, the cost of a T.L.P. was about the same as a conventional fixed steel platform and divers can be used, if needed, at these depths. The Hutton field also exhibits a low gas/oil ratio, so significant volumes of associated gas and gas liquids will not need to be processed, resulting in a reduced platform equipment payload.

The proposal is a significant development for tension leg technology and will be watched with great interest by the oil industry. Partners with Conoco in the Hutton venture are, BNOC, Gulf, Amoco, Amerada, British Gas, Mobil and North Sea Incorporated. The main design contract has been awarded to Brown & Root (UK) in association with Vickers Offshore (a subsidiary of British Shipbuilders). Mooring system design will be a subcontract.

3.2.7 Submerged Production System (S.P.S.) Exxon

The S.P.S. system is a full capability production system and consists of eight subsystems; template, drilling and completion, production manifold, remote control and safety shut-in, subsea pumps and separator, pipeline and pipeline connectors, production riser and floating facility, and maintenance facility. The template unit is the major component of the system (Fig. 3.2.7).[43]

The fluids produced by the wells are gathered by the production manifold on the template and routed by pipelines and an articulated production riser to a surface processing facility for storage and disposal. A drillship is used to install the template and for drilling the development wells. Surface controlled, electrohydraulic

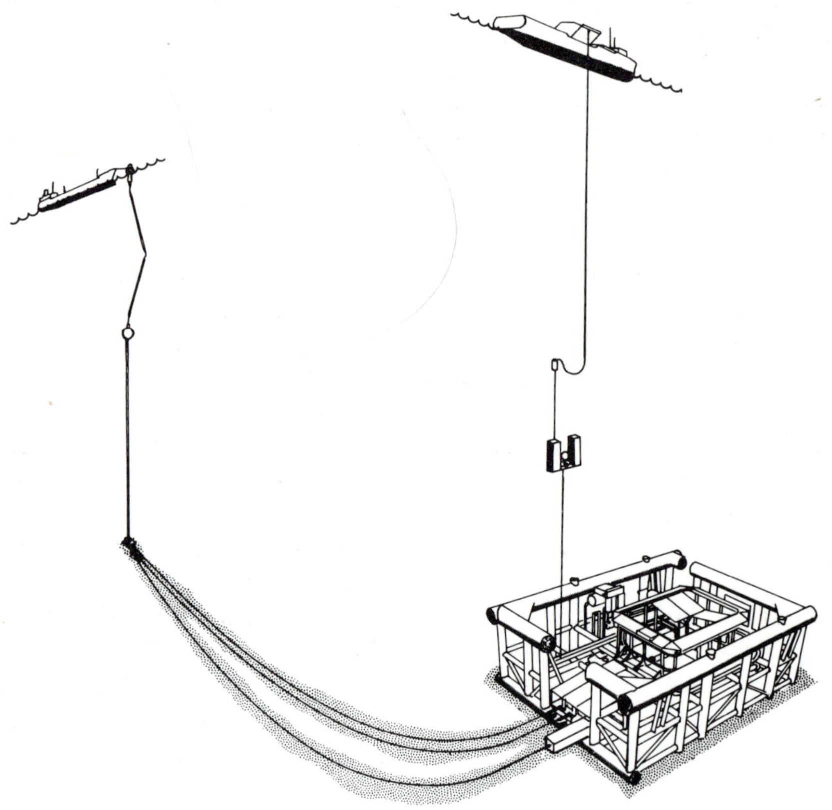

Figure 3.2.7 Submerged production system (Exxon S.P.S.).

supervisory control equipment is used to remotely control and monitor the ocean floor equipment and this eliminates the need to expose personnel to the ocean floor environment during installation, operation, maintenance and recovery of subsea equipment.

The manifold subsystem includes all subsea production gathering and water and gas injection equipment, arranged in a configuration required for manipulator maintenance. The manifold is a five header arrangement, with each header connected via an individual pipeline to the surface facility. Two headers are used for pump down tools or other well servicing tasks and the other three of larger diameter (12 inch and 6 inch) are generally used for production, gas lift and for injecting water or gas to maintain reservoir pressure. If production warrants it all these three headers can be used for production. A pump/separator unit can be included in the system to increase well rates and overall field production by lowering back pressure on the reservoir. The produced gas/liquid stream is separated and the separator pressure is employed to flow gas to surface facilities, while a bank of modular pumps transfers liquids.

A pollution control subsystem is incorporated and sensors will shut-in wells and automatically dispose of contaminants. The remote control system provides the command and power link from the surface control centre to all remote functions on the ocean floor. It is designed to provide fail safe operations in conjunction with safety alarm and shut-in systems.

The maintenance manipulator system is unique to this system and allows servicing of subsea equipment by removing and installing specially designed valves and control modules, thereby allowing replacement of malfunctioning components. An unmanned manipulator hauls itself down onto a cogged track on the template and is guided by C.C.T.V.; it can replace control valves and control modules subsea without any auxiliary equipment. All subsea equipment is designed to be depth insensitive and fully capable of operating at ambient pressure. Seawater sensitive devices are isolated from seawater by immersion in oil kept at ambient sea pressure by means of a pressure compensating device.

The system has been pilot tested in offshore conditions and has shown that deep water installation techniques are practicable and the maintenance system is competent to repair any faults likely to develop in the system. The pilot test system was designed for operational depths of 660 m and may be extended deeper with minor modifications.

Shell/Esso's recent intention to use a sophisticated Underwater Manifold Centre in the North Sea has its origins firmly based in the S.P.S. system. The U.M.C. was to be installed in 150 m water depth and connected to a fixed platform several miles away,[44] but its installation has now been delayed until 1982.

3.3 FUTURE PRODUCTION SYSTEMS

These systems incorporate concepts that are under serious consideration but are at an early stage of development. The minimum requirement for a system to be included would have involved a feasibility study, although detailed design studies or model tests may have been undertaken on the total system or system components.

3.3.1 Arcolprod: Articulating Buoyant Column Taylor Woodrow supported by the E.E.C. and the Department of Energy

This system is an advanced application of the principle of articulated buoyant columns, which are compliant with the action of waves, currents and wind forces. The facility will incorporate drilling and production workover and water injection and gas re-injection. Present studies are centred on a depth range of 200 - 400 m, although other studies are expected for applications in deeper and shallower depths.[45]

The configuration consists of a cellular concrete base anchored to the seabed and jointed by a flexible non-mechanical hinge to the foot of a prestressed concrete cylindrical column, which extends above sea level and accommodates drilling/workover rig, process plant, living quarters etc. (Fig. 3.3.1). The cellular base can house well-heads and manifolds in a 1 atms environment. Direct access is through a flexible tube at the foot of the column. The non-mechanical flexible joint consists of an array of flexible tendons made from a synthetic material (parafil) of high strength, to accommodate a wide range of motions and large buoyancy forces. The tendons will be protected and monitored and can be replaced individually if required.[46]

The column tapers from 35 m at the surface to 10.5 m (outside diameter) at the foot and the lower part of the column acts as a buffer store for oil production (60,000 barrels). Within the column a 5 m diameter sealed access shaft carries an elevator, goods hoist and safety ladder together with pipework and ventilation services from the upper deck to the base. The total well capability is for 30 well slots, additional satellite wells being provided for water/gas injection as necessary.

The system is very adaptable as the individual components can be employed separately or collectively in a range of offshore applications. Arcolprod can be designed for full production and drilling, production and workover, production only and to fulfil

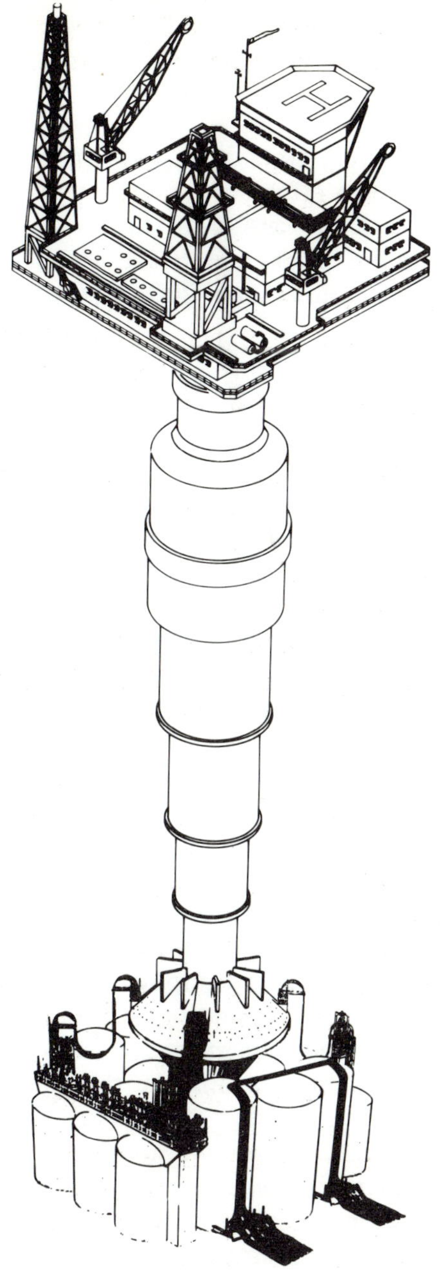

Figure 3.3.1 Arcolprod (Taylor Woodrow).

specific requirements as an oil storage structure. Simplified cellular bases could act as templates for predrilling wells or as chambers for subsea wellheads. The flexible joint can be designed for use in existing applications of compliant towers such as flare stack, mooring etc.

Hydrodynamic model tests have showed that in a variety of loading conditions, the effects of movement and acceleration on equipment and personnel are more favourable than large semi-submersibles; notably, pitch and heave are virtually eliminated. Present depth applications are 200 - 400 m although deeper systems may be feasible.

3.3.2 Under Seabed Tunnels for Oil Exploitation University of Newcastle

An interim report was prepared by Prof. Potts of the University of Newcastle for the Department of Energy into the feasibility of developing offshore oilfields by means of tunnels and underground excavations. The concept envisages drilling twin 5 m diameter tunnels from a land base up to 100 km offshore. The tunnels would rise gently to the destination above the well site 100 m below the rockhead. One tunnel would provide the access tunnel carrying fresh air, power cables and transit system and the other the return air-way and product pipelines. The tunnels would be constructed 30 m apart and cross drives would be constructed every 1000 m to facilitate cross-ventilation.[47]

Full face drilling machines and proven mining technology would be used to give tunnelling rates of approximately 1 mile per month. At the end of the tunnels would be a complex of excavations containing drilling equipment to bore the wells, well-head equipment and any separation and production equipment required. The tunnels could be extended to new sites determined by the limits of deviation drilling at the initial site and other prospects.

The study showed that there was no major impediment to such a proposal and that under certain conditions of water depth and distance from shore the proposed operation may be cheaper than present oil exploitation methods. The containment of a blowout in this concept may, however, be a problem.

Costs were estimated to be of the order of £1million per mile for tunnelling, with a proposal for a twin tunnel 25 miles offshore costed at £65million.

There is no specific depth limitation inherent in this concept, however it is probably unlikely that oil discoveries in deepwater will be sufficiently close to shore to make this method economically feasible.

3.3.3 Deep Sea Production Systems (D.S.P.S.)

Consortium: Sir Robert McAlpine & Sons
Humphreys & Glasgow
Rolls Royce Limited
BICC Limited
and E.E.C.
Department of Energy

The Consortium was formed to develop viable methods of producing oil from the continental margins in water depths from 200 to over 1000 m. The aim of the concept is to minimise the problems associated with deepwater and surface environmental conditions, by housing the majority of the production system on the seabed in large concrete modules in a dry one atmosphere environment. In this way the problems of high pressure risers for oil, water and gas are eliminated and the effects of surface conditions and water depth reduced.[48]

One type of system configuration would consist of subsea wellheads, flowline manifold, production modules, seabed flowlines, flare tower and tanker loading and power generation tower, supported by a vessel operating an access submersible. The novel elements of the system are the five seabed production modules, each 12 m in diameter and 65 m long, which provide a similar volume to a surface platform to house conventional plant and equipment. Each module contains a specific function of the overall system i.e. Habitat and Control, Water Injection, Gas Compression, Separation or Manifolding. The modules are provided with a separate fresh air and nitrogen supply from the tanker loading tower and respectively provide life support and an inert environment for the safety of processing equipment. Figure 3.3.3 shows that the habitat and water injection modules are normally an air environment, while the manifold, gas compression and separation modules are inerted. The latter can, however, be ventilated with air for major maintenance programmes.

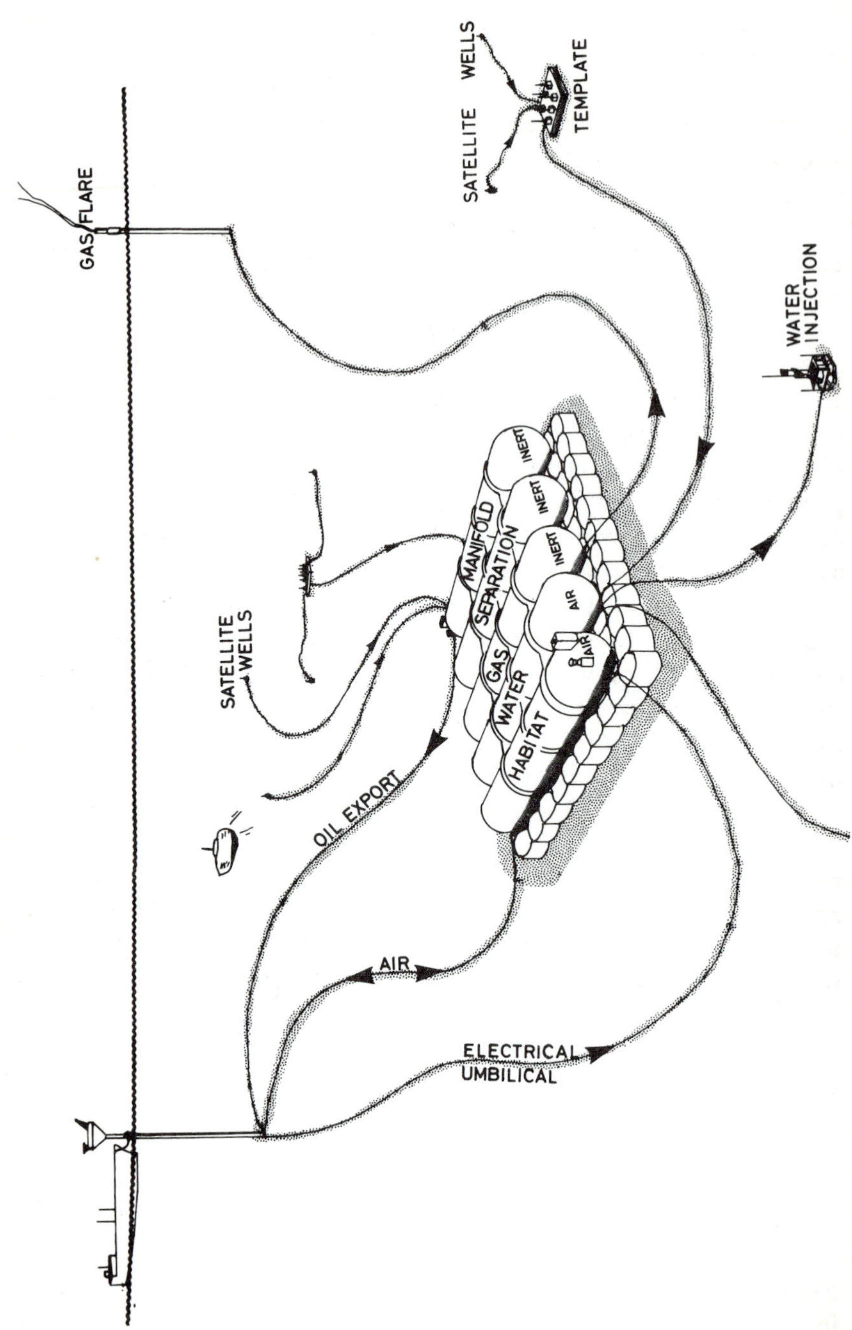
SATELLITE WELLS
TEMPLATE
GAS FLARE
WATER INJECTION
INERT
INERT
INERT
AIR
AIR
MANIFOLD
SEPARATION
GAS
WATER
HABITAT
SATELLITE WELLS
OIL EXPORT
AIR
ELECTRICAL UMBILICAL

All the oil processing equipment will be conventional surface systems, but remotely controlled and monitored from the computerized control room on the seabed, with repeat facilities on the surface, to increase reliability and reduce manning requirements. The modules will probably be constructed of reinforced concrete for depths up to 500 m; beyond this depth a composite sandwich construction of steel and concrete may provide the optimum method of combining the best properties of both materials to provide suitable characteristics for deeper units.

The complex will be manned with up to 50 personnel on a two week, 12 hour shift basis and access will be provided by a submersible from a surface support vessel. A special materials bell is to be used for transfer of equipment. Various safety and escape systems are incorporated in the design. Safe havens are provided for personnel awaiting rescue and self-contained escape capsules are available should environmental conditions restrict submersible operations.

The surface support facilities pose the major restrictions to the concept. Present designs of loading towers are not capable of dealing with the water depths, loading requirements and the difficulties of umbilical connections. The launching and retrieval of the submersible from the surface support vessel will be very dependent on environmental conditions.

The concept attempts to minimize the problems associated with deepwater and surface environmental conditions by housing the majority of the equipment on the seabed in a one atmosphere environment. The requirement for surface support facilities for power generation, ventilation and submersible launching and retrieval at present limit the advantages of the subsea production concept. Initial development of the system is likely to be based on the use of smaller modules for flowline manifolding or primary separation to form a development base for future work. The concept offers application in water depths of 200 to over 1000 m.

3.4 SUMMARY OF PRODUCTION SYSTEMS

Fixed structures of steel or concrete have established a foothold in shallow water and are likely to retain application in water depths to **200** m in the North Sea and **300** m in areas such as the

Gulf of Mexico. The Exxon guyed tower, which can be essentially considered as a hybrid fixed platform for drilling and production may play an important role in **300 - 600** m depth interval.

Floating production systems (F.P.S.) offer the potential of early production, deepwater application and marginal field development. The concept has minimum sensitivity to water depth and is relatively mobile. Operational experience on the Argyll field in the North Sea has established F.P.S. as a viable production method.

There are several versions of the tension leg concept of floating production systems and considerable development work has been undertaken on some of the systems. The tethered buoyant platform (T.B.P.) of B.P. - Vickers shows considerable promise: it is used for production only, the wells being pre-drilled before the production platform arrives. At depths less than 100 m, sway (surge) becomes significant and group tether load increases significantly and it is therefore unsuitable for shallow waters. Most problems have been solved for harsh environments to **450** m, although beyond these depths there may be problems with dynamic instabilities and riser design. Gulf Oil's recent feasibility study of a T.B.P. system, in which each seabed anchor is an eight well template claims application in water depths of **400 - 1000** m. Deep Oil Technology have tested a one third scale model of their three-legged system and are to simulate conditions for commercial platforms in water depths of **210** m.

The V.M.P. (Amoco) concept supports both drilling and production operations and has involved detailed design studies. The integral structural riser being a unique and a key element of the system. The wells are drilled and produced through the structural risers and the operation of the V.M.P. is essentially the same as a fixed platform. The Blow Out Preventer (B.O.P.) stack and Christmas trees are located on deck. The conductor and drive pipes acting basically as grouted tension piles. Detailed studies indicate that the system has flexible and practical application to water depths of **250 - 1000** m.

The Conoco tension leg design utilizes 'drillstring' type tethers and special flex joints to connect to the platform and to the subsea anchor plates. A separate subsea drilling template is used for production at the centre of the platform. The design concept is claimed to have application in water depths far in excess of that possible with existing platforms. The proposal to use the

Conoco concept on the Hutton Field (147 m depth) in the North Sea represents a significant development for tension leg technology and will provide the first practical operating experience of such systems.

The Exxon Submerged Production System (S.P.S.) is a totally different concept and can be considered as a subsea production system. A drillship is used initially to install a template and to drill the wells. The associated equipment then installed is a full capability production system remotely controlled and maintained from the surface using electrohydraulic techniques and a maintenance manipulator. Pilot tests offshore have shown that deepwater installation techniques are practical and the maintenance system is competent to repair any faults likely to develop in the system. It is designed for 660 m water depth, but could be extended easily.

The articulated buoyant column (Arcolprod) incorporates drilling and production facilities and allows access to the wellheads at one atmosphere via a shaft in the column from the surface to the seabed. The compliant action of the tower has good hydrodynamic characteristics and the components of the total system could be used independently for other functions. Design depth range at present is 200 - 400 m, although deeper systems may be possible. The application of the tunnel concept will probably not be suitable for deepwater developments offshore.

The Deep Sea Production System (D.S.P.S.) attempts to minimize the problem of deepwater and surface environmental conditions, by placing the majority of the production system on the seabed in dry one atmosphere modules. Each module contains a specific function of the overall system. The complex is manned and access is provided for personnel by using a submersible. Materials are transferred by a bell and various safety and escape systems are incorporated in the design. The surface support facilities pose the major restriction to the concept, although it offers application in water depths of 200 m to over 1000 m. Detailed feasibility studies have been undertaken (Table 3.4.1).

3.5 PRODUCTION SYSTEMS: CONCLUSION

The guyed tower and tension leg platform and their associated

TABLE 3.4.1 Summary of offshore production systems.

Type	*System*	*Sponsor*	*Depth (m)* *min.*	*max.*	*State of Art*
FIXED	Steel/Concrete	Various	*	300	Established
	Steel-guyed Tower	Exxon	280	600	Detailed Design 1/5 Scale Test. (Gulf of Mexico = 300 m Lena Field 1983)
FLOATING	Semi-sub	Advanced Production Systems	80	500	Detailed Studies
	T.B.P.	B.P. Vickers	100	600	Detailed Studies
	T.B.P.	Gulf-Oil	200	1000	Feasibility Study
	D.P.X.1	Fluor Subsea Services Deep Oil Technology	210	600	Model Tests 1/3-Scale
	V.M.P.	Amoco	330	1000	Detailed Design
	T.L.P.	Conoco	147	600	Design for Hutton Field North Sea — 1983
SUBSEA	S.P.S.	Exxon	*	660+	Pilot tests offshore. UMC Shell/Esso North Sea 1982.
	D.S.P.S.	D.S.P.S. Consortium	200	1000+	Feasibility Study
TOWER	Arcolprod	Taylor Woodrow	200	400	Feasibility Study
TUNNEL		Newcastle University	*	*	Feasibility Study

*No theoretical limits

components have been tested to the extent that there is little doubt that their full scale prototypes will be successful. The guyed tower will probably play an important role in 300 - 600 m depth range with the possibility of the tension leg concept extending to 1000 m. The application of the tension leg designs beyond 500 m water depth will be critically dependent on advances in the technology of riser design, mooring systems and diverless subsea systems. The problems encountered at shallower depths will be compounded by increased depth requirements and the possibility of dynamic instabilities.

The only designs suitable for very deepwater work are the Exxon S.P.S. system and the D.S.P.S. system. The highly automated S.P.S. system has been prototype tested successfully, but still relies on a high degree of surface support. The D.S.P.S. system has undergone detailed feasibility studies and offers a subsea production facility with conventional equipment housed in manned units on the seabed with limited surface support. The choice between the two systems revolves around the confidence that can be placed in fully automatic systems and the degree of surface support that is acceptable. Both systems should have application in shallower water depths once the technology has been established.

The philosophy of the D.S.P.S. does attempt to eliminate the problems of high pressure risers for gas and oil, and to minimize the problems associated with deepwater and surface environmental conditions. However, to achieve maximum benefit from the concept dependence on surface systems should be reduced. Surface interface equipment should only be incorporated when essential, and if alternative subsea systems are unlikely to become available.

Other factors that indicate that further investigation of the D.S.P.S. concept is justifiable:

1. The only system at present offering applications in water depths deeper than 1000 m, although S.P.S. may be extended to greater depths.
2. General application of the module concept for other intermediate areas of subsea production, i.e. manifold, primary separation, storage.
3. Potential application in ice infested regions, where ice scour

is not a problem, and busy shipping areas.

4. Use of the technology for exploitation of the oceans in other roles, i.e. coal mining, defence, etc.
5. Limitations of alternative technologies.
   (a) High cost of fully automatic subsea production systems and their dependence on surface support facilities.
   (b) Problems of T.B.P.
      (i) Static and dynamic excursion of platform
      (ii) Riser design
      (iii) Deepwater mooring problems
      (iv) Possibility of dynamic instabilities in depths greater than 500 m.
      (v) Sensitivity of platform to loading.
      (vi) Lack of diver intervention to subsea systems in deepwater.

There would seem reasonable cause to investigate the potential of the D.S.P.S. concept in further detail on the basis of present oil production systems proposed and tested.

4

OVERVIEW

The following major conclusions can be drawn from the previous analysis:

1. Significant prospects of finding commercially exploitable deposits of oil and gas offshore exist on the continental margins of the Atlantic, Indian and Arctic Oceans. The deposits will be difficult and expensive to find and extract and may involve operations in water depths approaching 2000 m by the end of the century.
2. The exploitation of resources in such deep waters will require new and expensive technology. The deepsea technology will probably advance by staged development, rather than by a technological breakthrough because of the high financial risks involved.
3. The capability to drill efficiently and safely over the upper continental slope and in relatively strong currents exists. Special deepwater drilling technology is evolving and should not be a restriction to the development of deepwater hydrocarbon deposits.
4. The riser system is at present the weak link between the sophisticated subsea and surface hardware. The extension of present riser technology for deepwater application will involve many complications and require considerable improvements in design.

5. The basic concept of subsea completions, manifolds and templates is well established. These systems will proliferate in existing production areas to depths of 300 m and developments during this period will form the technological base for future subsea activities in deeper waters.
6. The selection and design of an offshore production system will be determined fundamentally by the environmental conditions, the characteristics of the well and the financial restraints on the production time scale. The worldwide variation in development parameters would suggest that any proposed production system could have potential application under specific conditions.
7. The guyed tower will have application in the 300 - 600 m water depth range. The tension leg concept may extend to operation to 1000 m, although beyond 500 m its application will be critically dependent on advances in the technology of riser design, mooring systems and diverless subsea systems.
8. At depths greater than 1000 m only the Deep Sea Production System (D.S.P.S. — McAlpine) and possibly the Submerged Production System (S.P.S. — Exxon) would appear to have the necessary potential for this application. The choice between the two systems will be determined by the confidence that can be placed in fully automatic systems and the degree of surface support that is acceptable. Both systems will have application to shallower water depths once the technology has been established.
9. In the specification of a suitable production system for various water depth ranges a contentious area exists between 500 - 1000 m. The use of total subsea production systems or surface piercing structures at these depths, is an area of uncertainty.
10. The potential of various production systems should be continuously re-assessed in the light of contemporary developments of the system to retain a correct perspective. Capital and running costs of the system will be a critical consideration.

The limitations of automatic subsea production systems and tension leg concepts for oil production indicate that a more detailed study of the manned subsea complex (D.S.P.S.) would be justified for water depth applications from 500 to 2000 m. The studies will also be of value in the general application of the subsea module concept for oceanographic exploitation.

REFERENCES

1. 'Future Offshore Exploration and Production Systems' by Dr. J. Birks, BP Limited, in *Techno-Economic Appraisal of Offshore Production of Oil and Gas*, Society of Underwater Technology, February 1978.
2. 'Hydrocarbon Potential of Deepwater' by H.R. Warman, in *Sea Floor Developments: Moving into Deep Water*, Royal Society, June 1977.
3. 'Introduction to Aspects of Economics and Logistics' by Dr. J. Birks, BP Limited, in *Sea Floor Developments: Moving into Deep Water*, Royal Society, June 1977.
4. 'New Slant on World Petroleum Resources' by P. Wood, *Ocean Industry*, p. 59 (April 1979).
5. *Britain's Oil in the Eighties* by P.B. Baxendell, Shell (UK), Scottish Institute of Petroleum, October 1978.
6. *The Offshore Energy Technology Board: Strategy for Research and Development*, Energy Project No.8, Department of Energy, 1976.
7. 'Grand Banks — Another North Sea', *Ocean Industry*, p. 7 (February 1980).
8. 'Deepwater Finds', *Noroil*, p. 19 (February 1980).
9. 'Future Depth Requirements for Underwater Tasks' by J.R. Brooks, UK Department of Energy, in *Alternatives to Divers in Deepwaters*, Society of Underwater Technology, November 1979.
10. 'Technological Aspects of Offshore Oil and Gas' by Prof. Bishop, University College, in *Deepwater Exploration and Development 10 years on*, Greenwich Forum, October 1978.
11. 'European Operators Spell Out New Technology Plans', *Offshore Services*, (July 1979).
12. *Techno-Economic Appraisal of Offshore Production of Oil and Gas*, Proceedings of Society of Underwater Technology Conference, February, 1978.
13. 'Implications of Deepsea Exploration and Development for the Development, Restructuring and Redeployment of British

Industry' by D.E. Lennard, in *Deepwater Exploration 10 years on*, Greenwich Forum, October 1978.

14. 'Techniques for Drilling Well in Deepest Water' by D. Smith, Offshore International S.A., *Ocean Industry*, p. 39 (June 1979).
15. 'Drilling in Deep Water, Strong Currents' by F.E. Shanks, D.S. Hemmett and H.L. Zirkyref, Sedco Inc., *Ocean Industry*, p. 347 (June 1979).
16. 'Recent Advances in Drilling Tools and Technology' by Scott Weeden, *Ocean Industry*, p. 40 (June 1979).
17. 'Trends in Offshore Drilling' by C.R. Palmer, A. McLenon, T.C. Rogers and J.G. Somells, *Ocean Industry*, p. 39 (January 1979).
18. 'Glomar Explorer to be Converted', *Sea Technology*, (December 1979).
19. 'Marine Risers the Weak Link in the Chain' by Derrick Booth, *Ocean Industry*, p. 38 (March 1979).
20. 'How to Measure the Growth of Offshore Technology — A Graphic Method' by Derrick Booth, *Offshore Services and Technology*, p. 20 (January 1980).
21. 'Offshore Subsea Engineering' by E.C. Goldman, Shell International, Le Hague, in *Sea Floor Developments: Moving into Deep Water*, Royal Society, June 1977.
22. 'Subsea Completions Increase North Sea Production' by David Buckman, *Ocean Industry*, p. 54 (July 1979).
23. 'Certification Practice — Subsea Completions', *Underwater System Design*, p. 24 (August 1979).
24. 'Subsea Production Systems', Society for Underwater Technology, *Offshore Engineer*, (February 1979).
25. 'Offshore Trials of Prototype Subsea Wellhead Chamber' by D.L. Hingston, Vickers-Interteck Limited, *Offshore Europe*, (September 1977).
26. 'Garoupa Subsea Production System' by J. Delong and S.M. Adamson, Lockheed Petroleum Services, *Offshore Europe*, Aberdeen (September 1979).
27. 'Subsea Wellhead Equipment — A Guide to Remote Control Systems' by J. Mason, *Offshore Services*, p. 25 (March 1979).
28. 'Management of Subsea Techniques for Offshore Oil and Gas Development' in Society of Underwater Technology Conference, Aberdeen, February 1980.
29. 'Concrete Platforms' by Dr. Jensson, Norwegian Contractors, in *North Sea Developments: Experiences and Challenges*, Institute of Petroleum, Glasgow, April 1979.

30. 'Shipbuilder and Offshore Diversification' by A. Scott-Belch, Scott-Lithgow, *North Sea Developments: Experiences and Challenges*, Institute of Petroleum, Glasgow, April 1979.
31. 'Floating or Fixed', *Noroil*, p. 102 (May 1980).
32. 'Floating Production Systems' by James Barrack and Rayburn Hamilton, Worley Engineering, European Offshore Petroleum Conference, October 1978.
33. 'Early Production Systems Using Floating Platforms' by J. Parker, Vickers-Aker Consortium, *North Sea Developments: Experiences and Challenges*, Institute of Petroleum, Glasgow, April 1979.
34. 'The Guyed Tower as a Platform for Integrated Drilling and Production Operations' by L. Finn, J. Wendell and T. Loftin, Exxon Production and Research Company, European Offshore Petroleum Conference, 1978.
35. 'Exxon Gives Data on Guyed Tower', *Ocean Industry*, (March 1980).
36. 'B.P. Development of Tethered Buoyant Platform Production System' by J. Laing and C. Hedley, BP Limited, *North Sea Developments: Experiences and Challenges*, Institute of Petroleum, Glasgow, April 1979.
37. 'Whats Ahead in Offshore Production Technology', *Ocean Industry*, p. 44.
38. *Technical Information — Gulf Tension Leg Platform.* Gulf Research and Development Company, Houston, 1979.
39. 'Tethered Buoyant Platforms: T.B. or Not T.B.', *Offshore Engineer*, p. 32 (February 1979).
40. 'The Vertically Moored Platform for Deepwater Drilling and Production' by M.Y. Berman and K.A. Blenkam, Amoco Production Company, O.T.C., Houston, May 1978.
41. 'Conoco Tension Leg Platform with Double Water Depth Capability' by D.M. Taylor, *Ocean Industry*, p. 35 (February 1980).
42. 'Hutton's T.L.P. — Conoco Reveals Shape of Things to Come', *Offshore Engineer*, (February 1980).
43. *Submerged Production System — A Final Report* by J. Burkhardt and T. Michie, Exxon Company, O.T.C., Houston, May 1979.
44. 'Shell/Esso U.M.C. — The Latest Word in Subsea Sophistication', *Offshore Engineer*, (September 1979).
45. *Arcolprod: Articulating Buoyant Column* Technical Notes from Taylor Woodrow Construction, September 1978.
46. '*Arcolprod:* TayWoods Buoyant Idea', *Offshore Engineer*,

p. 87 (September 1979).

47. *An Investigation into the Feasibility of Developing Offshore Oilfields by Means of Tunnels and Underground Excavations* by Professor Potts and Mr Walton, University of Newcastle for the Department of Energy.
48. *Seabed Hydrocarbon Production — Phase I Summary Report* by Deep Sea Production Systems (Sir Robert McAlpine, Humphreys and Glasgow, Rolls Royce, BICC Limited), July 1979.
49. *Total Subsea Production System (TSPS) Subseamac One*, Sir Robert McAlpine and Sons Limited, circa 1977.
50. 'Deep-Sea Well Readied for La. Gulf', *The States Item*, New Orleans, 8 February 1980.

SECTION B

The Manned One Atmosphere Underwater System

5
INTRODUCTION

The concept of 'inner space' as a frontier for the exploration and exploitation of the deep oceanic areas is an appropriate comparison to outer space. The reasons which make outer space a challenge, i.e. adventure, scientific curiosity, military and commercial exploitation and the consequent need for technological advancement equally apply to the deep ocean. The presence of the high hydrostatic pressures and a corrosive marine environment in fact make the conditions more severe and the challenge more exacting. The similarities are accentuated by the requirements for manned operations in either of the deep spaces. The primary consideration in any manned operation is the safety of the personnel in the face of the environment, which in the case of exposure can at once be fatal and exclude any chance of survival.

One implication of the common characteristics of inner and outer space was the natural diversification of aerospace companies in the United States into the oceanic field in the late sixties with the demise of the outer space program i.e. Lockheed, Rockwell, Westinghouse. The result being complex projects, such as the Sealab and Tektite hyperbaric systems and detailed feasibility and design studies of one atmosphere manned underwater stations (M.U.S.) for the U.S. Naval Engineering Facilities Command. The initial impetus of this work has now subsided and present habitat activity worldwide is at a very low level, except for the use of hyperbaric habitats for commercial saturation diving operations.

The ever-present dangers inherent in diving systems are well known. Present opinion amongst authorities aims for replacement of the diver in subsea work even at shallow depths. Physiologically, the maximum diving depth is estimated to be of the order of 500–700 m. A practical working depth for the diver will be even less, something of the order of 300 m. The limited work tasks that can be undertaken in deep waters and the expense of such operations will restrict the use of hyperbaric methods for oceanic exploitation.

To expect man to work for considerable periods in the ocean depths, it would appear fundamental to provide him with a safe working environment and be able to assure him of his survival. Only by protecting him from the severe hydrostatic pressure and creating an artificial terrestrial environment can we begin to approach his natural working conditions. To assure him of his safety he must have confidence in the system and be sure of a safe return to the surface or to a land base.

There exists considerable experience in the design of one atmosphere manned underwater structures, from the worldwide use of military submarines. The nuclear submarine programme with the requirement for long submergence times is especially relevant to our studies.

The following technical study will therefore draw on the technology of nuclear submarines, space systems and other related areas to attempt to identify suitable technologies for incorporation in the concept of the one atmosphere underwater system. This section provides a multi-disciplinary investigation into the relevant fields of interest. Potential developments and areas requiring further detailed study are identified.

The following topics will be investigated:

(a) Brief survey of past 1-atmosphere habitat designs
(b) Environmental Control and Life Support
(c) Physiological Implications of Atmosphere Control
(d) Human Factors
(e) Psychological Aspects
(f) Submersibles
(g) Escape and Rescue
(h) Subsea Power Supplies
(i) Safety

The area that at present holds the greatest potential for the application of the one atmosphere manned underwater structure is the subsea production of oil (Fig. 5a and b). The information base is therefore assessed with a view to identifying possible operating systems for this application. However, generally the technology will be suitable for many other oceanic applications.

5.1 PAST HABITAT DESIGNS

A brief review of past habitat systems would appear appropriate to establish the present information base. The approach distinguishes between general habitat projects and those directly related

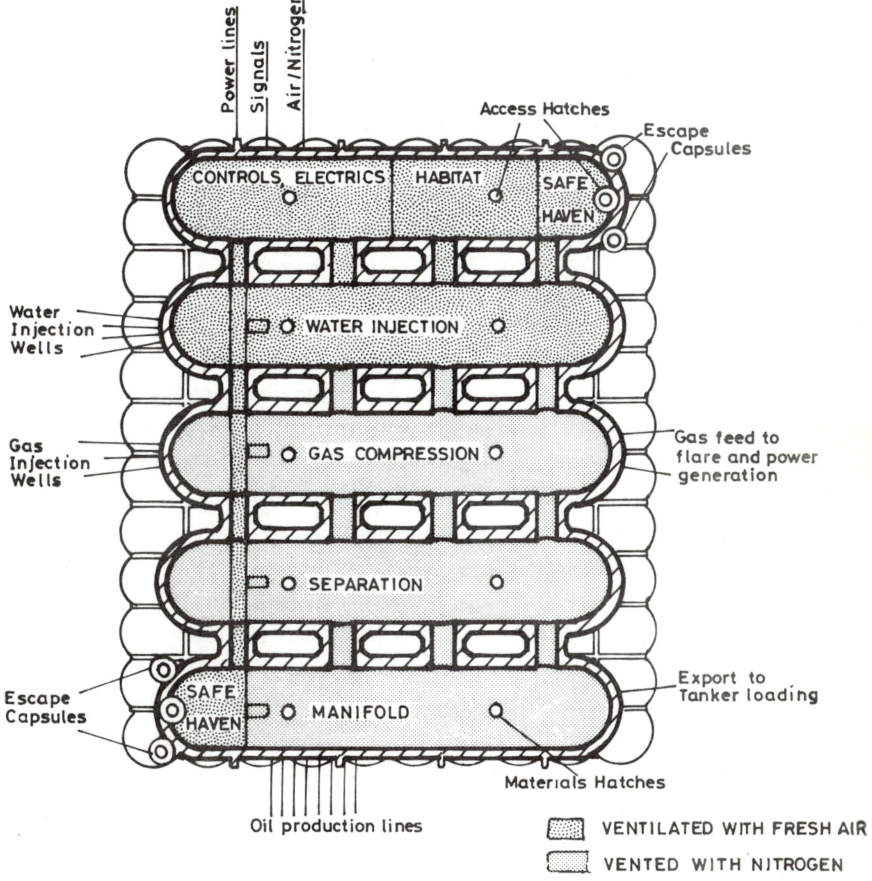

Figure 5(a) Typical plan of module layout — 100,000bpd.

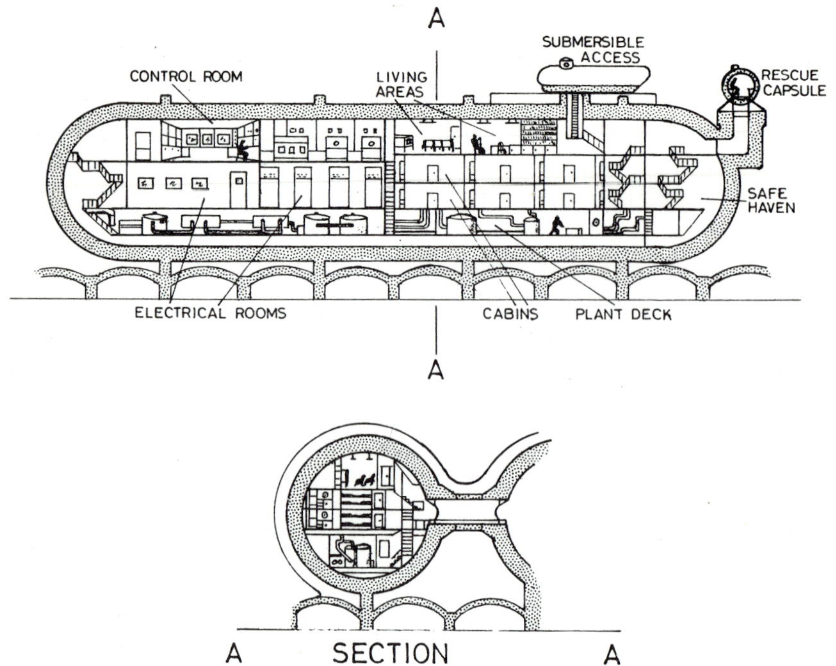

Figure 5(b) Typical section of habitat and control module (D.S.P.S.).

to the oil industry. The clear distinction between hyperbaric habitat systems and one atmosphere habitats is invoked to select only studies relevant to the one atmosphere concept, although the indirect value of hyperbaric programmes is appreciated. For the purpose of this book a habitat is defined as a structure which encloses a one atmosphere environment in which man can operate. The pressure hull for our specific application may be assumed to be located on the ocean floor.

The results of computerized literature searches show the large number of design studies carried out in the sixties and the dearth of information generated during the last decade. More recent relevant information is available from the military submarine programme, commercial submersible operations, space and the use of one atmosphere completion and manifold centres for sub-sea oil production by CanOcean Resources Ltd, Canada.

The United States Naval Facilities Engineering Command-Ocean Engineering is concerned with the construction of under-water structures. Reference 51 gives a very good overview of underwater structural concepts under development during the sixties. The paper deals with three main categories, above-bottom, on-bottom, and in-bottom structures. The above-bottom concept deals with platforms, underwater storage systems etc. and the in-bottom system with tunnels and complexes under the seabed. The review of the on-bottom structures identifies those detailed design studies of one atmosphere manned underwater structures sponsored by the Naval Civil Engineering Laboratory in 1967. These studies are of major interest to the project and were carried out by Westinghouse Electric, General Dynamics and the South West Research Institute, under the general title of Manned Undersea Station (M.U.S.).

The conceptual studies explore the limits of various design problems, placement and recovery methods and other operational capabilities. The stations are designed to support a five man crew at atmospheric pressure for an indefinite period at 6000 ft water depth. The crew is replaced and re-supplied every 30 days, with provision for emergency life support for an additional five days. The pressure hull design of each structure was constrained to a single geometry configuration which should be inherently positively buoyant.

These three studies were essentially the only detailed reports on one atmosphere manned underwater structures that were identified in the literature search, apart from related items in other fields. The basic features of the three M.U.S. designs are outlined.

5.1.1 Toroidal Manned Underwater Station. Westinghouse Electric (Fig. 5.1.1)

The pressure hull is a toroid of 40 ft overall diameter and tube diameter of 10 ft. The shell is constructed of HY-140 steel and is internally ring stiffened. A hemispherical foundation is used as a method of providing an easy levelling system by means of a ball-and-socket action with the toroidal hull. The system capacity can be enlarged by mating similar hulls on a vertical stack.

The power supply is an isotopic heat source in combination with a steam Rankine turbo-electric power generation system (30 kW) to supply minimum power for continuous station requirements and life support. A reserve battery system handles station power peaking demand.

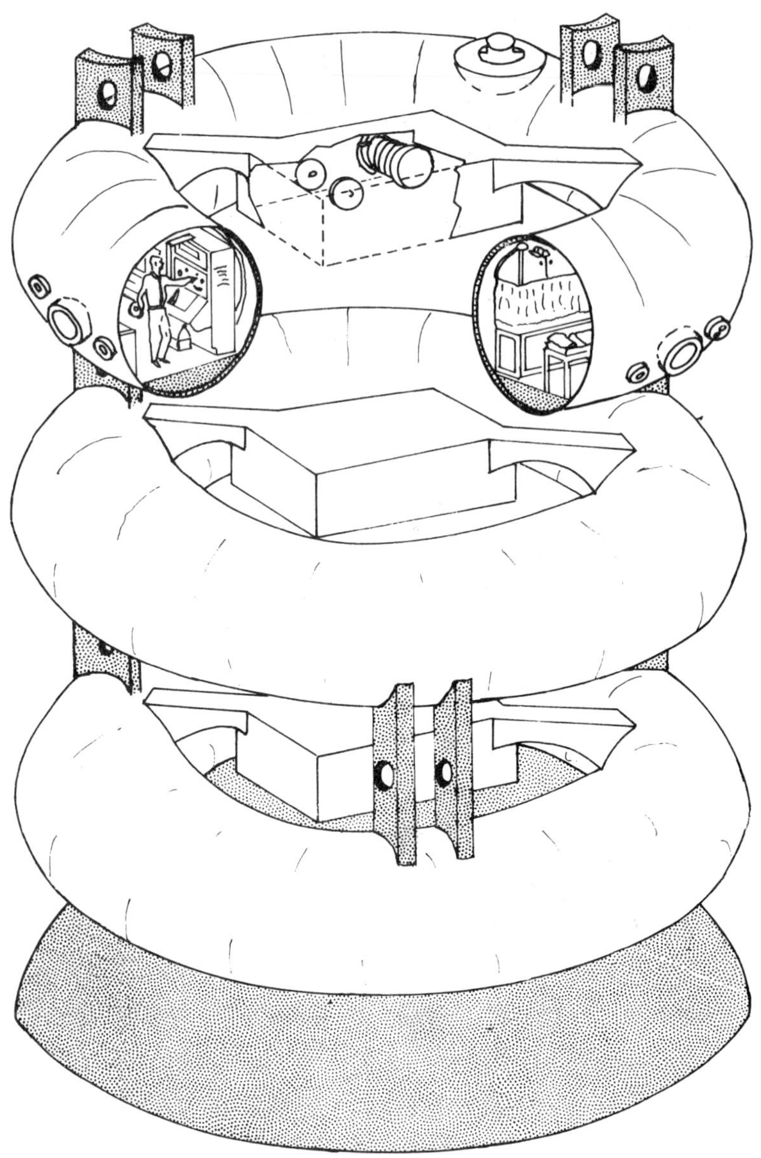

Figure 5.1.1 Toroidal manned underwater station (Westinghouse Electric).

Atmosphere control is through temperature and humidity control, carbon dioxide scrubbing, oxygen generation and CO/H_2 burner plus precipitator with the associated instrumentation and equipment. Crew interchange is suggested for long missions at 30 day intervals by using a submersible for support *in-situ.*[52]

5.1.2 Cylindrical Manned Underwater Station. General Dynamics/Electric Boat (Fig. 5.1.2)

The pressure hull consists of a vertical hemispherically encapped cylindrical steel hull of HY-130 steel, with two spheres attached to

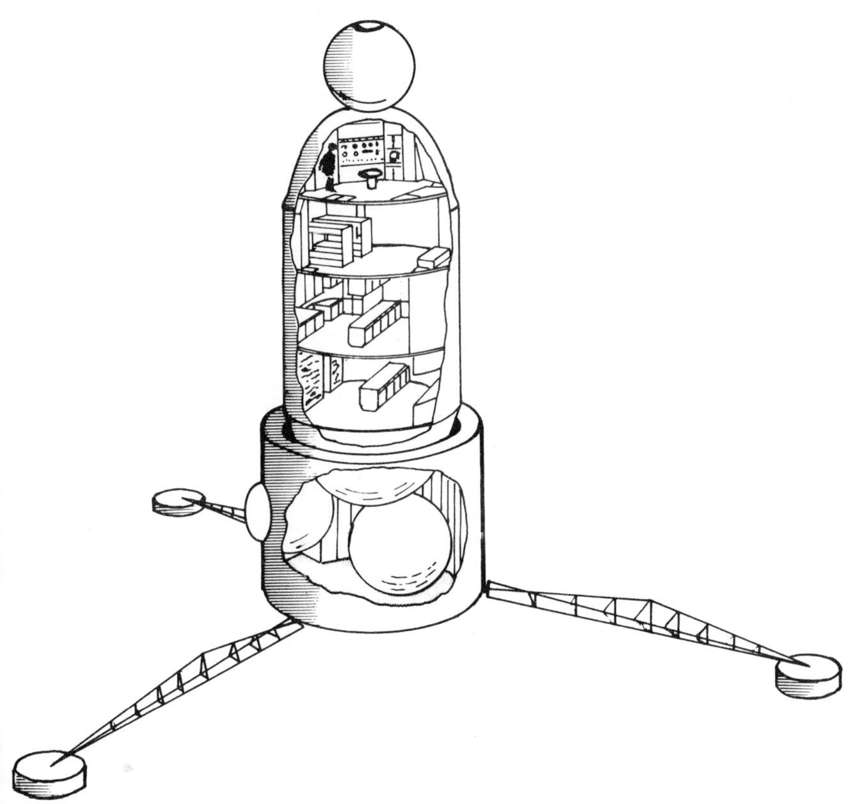

Figure 5.1.2 Cylindrical manned underwater station (General Dynamics/Electric Boat).

the cylinder, one for observation and the other for access. The hull's overall length is 51 ft and the habitat diameter is 12.5 ft and both of the attached spheres are 8 ft in diameter. A radioisotopic power system is contained in a vertical 9-ft diameter cylinder to supply the maximum power requirement of 20.5 kW. A reactor powerplant was also considered.

A ballast control system is utilized for free ascent and descent and tripod landing gear is used for the foundation system, which can also accommodate any bottom slope.

Oxygen storage was recommended, with CO_2 removal, CO/H_2 combustion and contamination control with appropriate instrumentation and equipment. Water is stored and condensation

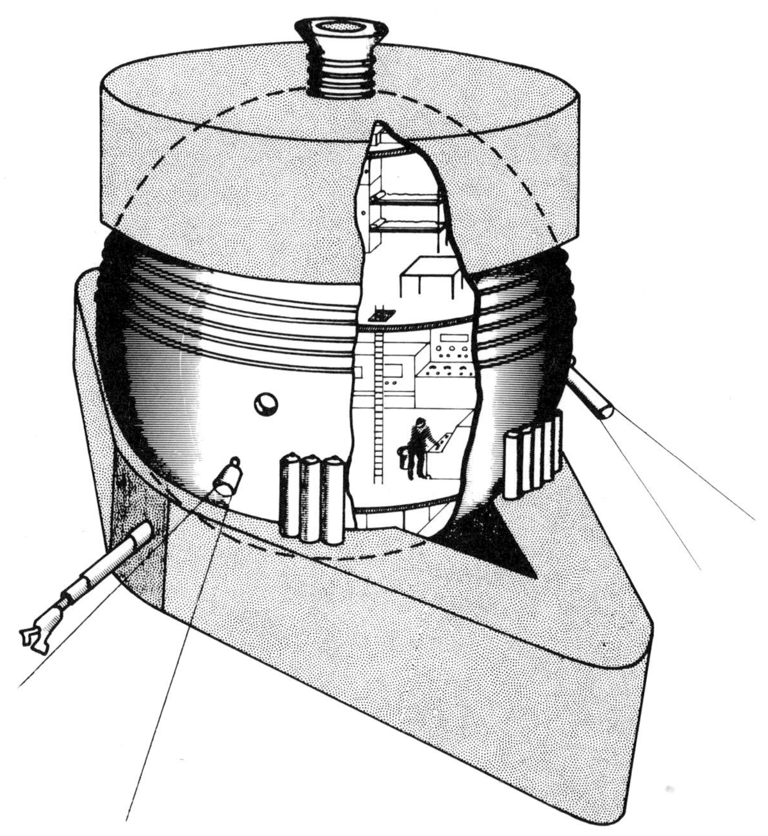

Figure 5.1.3 Spherical manned underwater station (Southwest Research Institute).

recovered for washing. Hull penetrations are restricted to inter-hull junctions and observation/access spheres. Larger capacity is given by multi-hull systems.[53]

5.1.3 Spherical Manned Underwater Station: Southwest Research Institute (Fig. 5.1.3)

The pressure hull is a 20 ft diameter HY-140 steel sphere. Oxygen and battery power are stored externally to the sphere. Descent and ascent are accomplished with an anchor and winch-down enplacement technique.

The required electric power load is 15 kW and atmosphere control is by CO_2 absorption and O_2 replenishment. Stored water is used with recovery of atmospheric water for washing.[54]

5.1.4 Study of One Atmosphere Manned Underwater Structures (SOAMUS)

These three basic conceptual studies were then used as the basis for an overall design study of a multi-parameter application of manned underwater structures. The study was prepared by the Ocean Systems Operations Division of North American Rockwell Corporation on behalf of the Naval Facilities Engineering Command in 1968. This very detailed study identifies roles and missions, considers structural concepts and subsystems and cost analyses possible systems for the combination of parameters shown in Table 5.1.4.[55]

In this book it is not possible to go into this work in detail, although much useful information will be gained from this work in other areas of investigation. It is, however, pertinent to list

TABLE 5.1.4 Study parameters for one atmosphere structures.

Number of Men	5, 25, 100, 250, and 1000
Underwater Depth (ft)	600, 200 and 6000
Duration (days)	30, 90, 365 and unlimited
Offshore Distance (miles)	100, 500, 1500 and 2500

some major conclusions from the study that are relevant to this application.

Findings and Conclusions

1. There appear to be no crucial technical feasibility problems, whose solutions are not expected to be fulfilled.
2. Umbilical systems will probably pre-empt inshore manned structures of long duration and intermediate to large manning levels, as these systems have an overwhelming support cost advantage.
3. Umbilical systems are expected to be feasible as extensions to present existing technology.
4. All geometries for pressure hulls; spherical, toroidal, cylindrical and prolate spheroidal appear feasible and offer multi-hull configurations.
5. There are difficulties with the fabrication and transport of large capacity single hull systems.
6. There are large station supply and support costs.
7. Nuclear powerplants are generally recommended as primary power sources for all non-umbilical structures regardless of size and mission duration. Small stored power units are recommended for emergency power. Feasibility is not a problem in power supplies.
8. Life support equipment for an underwater structure is basically dependent on manning level and mission duration. Feasibility is not a problem, although continued development is required.

Costs

9. Personnel and support costs are considerably larger than the initial investment. Typically two thirds of the total system costs are allocated to personnel and support costs.
10. Using cylindrical hull shapes for representative calculations, spheres were less expensive at shallower depths, but nearly identical at 6000 ft. Prolate spheroidal shapes were essentially the same as cylindrical in terms of cost. Toroidal shapes were consistently more expensive.
11. In small structures costs are dominated by powerplant costs.

Site-based charges decrease with manning level and hull-based costs increase with manning level producing an investment cost minimum. This suggests the possibility of a minimum cost multiple hull station that has elements of about 50 men size for 6000 ft depths and larger sizes for shallower depths.

12. The vertical transfer of personnel, equipment and supplies is a very large contributor to *in-situ* expenses and a trade off can exist between the installation of an umbilical and surface support, as a function of manning level and distance from shore.
13. In crew sizes greater than 25 men the powerplant capacity can be increased without prohibitive increases in cost as nuclear power plants recommended for this use decrease in specific power cost with increase in power level.
14. The difference in cost between base maintenance of a ship and the *in-situ* maintenance of a manned underwater structure was, surprisingly minor.

These reports contain a wealth of information on the design and operation of manned one atmosphere systems which will be drawn on in detail for the specific areas to be investigated further. The age of the publications will obviously require an attempt to establish the present state of the art in many of the subsystems discussed. The conclusions drawn will have relevance to various system concepts proposed for the exploitation of oil in deepwaters by the use of manned underwater structures.

N.B. No information was available on other relevant systems, i.e.

(a) Ocean Systems — North American Rockwell Corporation.
NAVSITE — 1 atms R & D 10 men at 1000 ft.
(b) Lockheed Missiles and Space Company.
Concept of M.U.S. 1 atms at 6000 ft 5 men 30 days. Above-bottom.
(c) General Dynamics
DOMAINS — 1 atms 5 men at 2000 - 6000 ft.

5.1.5 One Atmosphere Structures in the Oil Industry

The present use of one atmosphere subsea structures in the oil industry is generally restricted to their use for subsea completion, manifolding and pipeline connection systems. CanOcean Resources Limited probably have the most experience of their use in this field to date and have recently installed a large subsea production complex in the Garoupa field, offshore Brazil for Petrobras.[26]

The one atmosphere chambers or cellars house conventional oil processing equipment on the sea floor and intervention to the cellar is normally by an umbilically connected service capsule with surface support. Technicians are able to install, maintain and repair equipment in a shirt-sleeve environment during temporary occupation of the cellar (Fig. 5.1.5a).[57]

The dry chambers are used to house wellheads, control systems, valves and associated piping. Flow line pull-in techniques allow rapid installation of subsea pipelines. Chambers are also used for manifolding, attachment of service and production risers, and welding of subsea pipeline, all at one atmosphere.

The wellhead cellar protects the wellhead and associated equipment from the salt water environment and ambient pressure and is usually attached to the casing profile by the drilling vessel before it leaves the site. The cellar can be used for oil or gas production, water or gas injection and pump down tools can be accommodated for well servicing and gas lift.

The manifold centre is used to commingle a number of subsea wells and reduce the number and length of flowlines. In the normal mode of operation all production and shutdown operations are automatically controlled and monitored from the surface control centre.

An electro-hydraulic system links the control centre to a minicomputer located in the manifold centre, which performs remote control and data acquisition. If control is lost from the surface, local supervisory control is maintained by the computer on batteries for up to 96 hours or until remote control is restored. If anomalies occur meanwhile, the computer performs an automatic shut-down. A breathable air supply is maintained by suitable ventilation to control the CO_2 level and breathing masks are supplied should air become contaminated. Analysis and detection systems are provided for CO_2 and O_2 levels, hydrocarbon explosive limits and H_2S. Halogen 1301 agent is used for fire extinguishing and activated automatically by hydrocarbon lower explosive limit analysers.

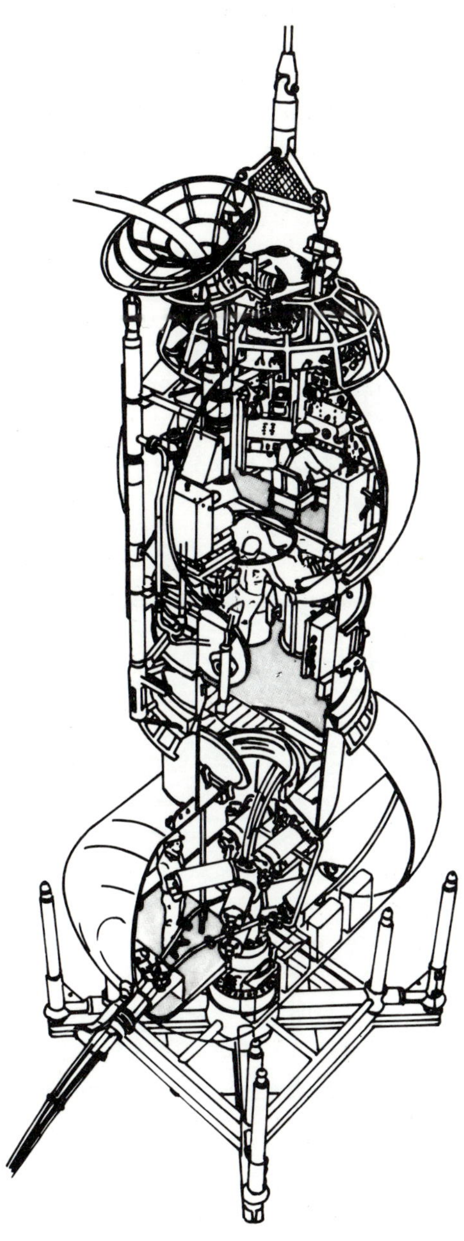

Figure 5.1.5(a) Wellhead cellar (CanOcean Resources).

Subsea one atmosphere chambers mounted at the base of a platform have been used to permit welding of connections between subsea pipelines and platform risers. The pipes are drawn in the chambers and a special jointing section is prepared on the surface after accurate measurement. The welding is done in a one atmosphere environment using a gas/metal arc process which produces

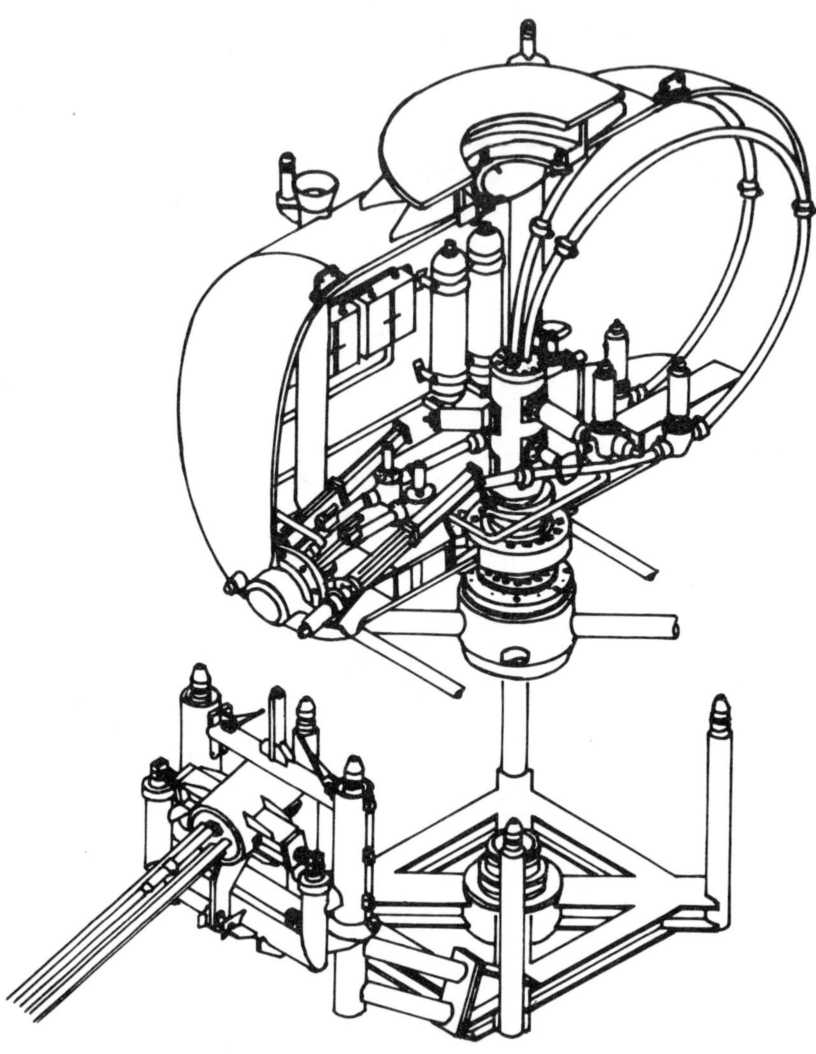

Figure 5.1.5(b) Wellhead cellar (Cameron).

little smoke; normal air circulation provided by the service capsule is sufficient to ventilate the cellar.

The service capsule, which is a buoyant steel capsule, winches itself down to mate with the cellar of the structure and carries a crew of 3/4 men. The umbilical connection to the surface support vessel is buoyant and carries air, power and communications. Voice communication and CCTV surveillance are maintained and emergency power and air supplies are available should the umbilical become broken. Instrumentation in the capsule gives air flow rates, CO_2 content, power levels, capsule pressure and water depth. Alarm systems monitor for small amounts of hydrocarbons and contaminants.

Cameron Iron Works are also developing a one atmosphere wellhead system which is serviced by a submersible (Fig. 5.1.5b).[58]

Application

In the International offshore oil industry at present there appears to be a firm trend towards 'WET' subsea systems i.e. those operated at ambient pressure by remote control, mainly because of the value in concentrating safety equipment below the mudline. Some consider that as a simple guide to the future we could expect single wells to be wet and multiwell systems to be dry.[24] In the move into deeper waters the proliferation of one atmosphere systems would appear necessary.

The cost of repair and maintenance of subsea equipment over a period of many years will be as important as the initial capital cost in choosing 'wet' or 'dry' subsea systems. Access to the facility must be provided no matter what maintenance procedure is adopted. Manipulators, although expensive, are useful for some routine wet servicing but are incapable of handling the unexpected problem. The diver is adaptable and can handle maintenance tasks requiring innovation, but has a depth limitation of 300 - 500 m and is also expensive. The useful combination of diver/technician required for production equipment maintenance will be limited and the work capability in deep waters restricted.

The encapsulated one atmosphere system offers the facility of greater working depths, an ideal working environment for non-diving personnel and relatively unrestricted bottom-time. Adequate servicing tools can be made available for use on quality surface equipment encapsulated in the chamber. A bonus feature of the encapsulated system is that the enclosures isolate potential sources of environmental pollution.

Although we are not specifically concerned with the relative merits of one atmosphere, neutrabaric or wet subsea systems there do appear to be operational advantages in using one atmosphere systems in deeper waters. The real value of the experience gained in the use of one atmosphere systems in oil related activities is that information will be available on atmospheric contaminants, pressure levels, leaks, explosive limits and support logistics that will be relevant to this book.

5.1.6 Submarines and Submersibles

The operational requirement in the design of conventional submarines and smaller submersible craft has involved the provision of suitable life support systems for normal and emergency conditions over relatively short time periods. The introduction of the nuclear submarine, with extended submerged patrol periods has involved a new level of sophistication in atmosphere control systems. The concept of the manned underwater structure is very similar to the nuclear submarine, but static rather than dynamic. Structural design, safety and rescue facilities will also be common features of each. Considerable programmes of research have been undertaken by the United States and United Kingdom Naval Authorities into the psychological and physiological implications of placing men in submerged enclosed environments for long periods.

The technologies of submarine and submersible design should therefore be investigated for details of structural design, safety, rescue and atmospheric control systems. Past research should indicate physiological and psychological problems that should be avoided or may result from our concept of a manned underwater structure.[60,61]

5.1.7 Space Systems

Space systems are of interest primarily as a data source for atmosphere control systems for life support. The payload penalty for space systems is critical, so high capital cost and better integrated life support systems are utilized in order to save weight along with exotic non-air-breathing power supplies. The level of cost that is acceptable exemplifies the distinction between space and subsea technology in these areas.

In association with the space programme much research has also been done on the physiological and psychological considerations involved in long-term space missions and is relevant to this work.

The high cost, sophisticated technology developed during the last two decades for the space programme should be investigated to determine the practicability of its application to the marine environment, especially in the areas of life support and non-air-breathing power sources. Human factors research should also be assimilated into the concept of the manned undersea structure.[62,63]

5.1.8 Hyperbaric Systems

These systems differ completely from the one atmosphere system because of their elevated pressure conditions and are therefore only related systems. Programmes such as Sealab and Tektite being characterized by relatively short mission times, small manning levels and shallow water depths.[64]

The main objectives of these types of programme have been heavily weighed with the human factor and performance measurements with external work limited to oceanographic observation. Materials, equipment and instrumentation development, however, has been an important part of the work. The considerable use of saturation diving techniques offshore commercially in the last decade has also led to the worldwide use of hyperbaric habitats.

The experience from hyperbaric work may offer useful information on: instrumentation for the measurement of atmosphere constituents and contaminants, physiology, psychology, human factors and breathing apparatus. This work may have more direct value if a requirement should exist for divers to exit from the structure at shallower depths for maintenance tasks on peripheral systems.

5.1.9 Other Related Fields

Technology from other related industries such as aerospace, shipbuilding, nuclear industry etc. may also offer significant contributions to the design and safe operation of the subsea complex.

6

ENVIRONMENTAL CONTROL AND LIFE SUPPORT

The provision of a suitable work environment within a one atmosphere underwater structure can be considered ideal if the conditions duplicate the average standard for the population working on the surface of the earth. The normal operating environment of such a system is, however, usually determined by establishing conditions within a range such that no physiological or performance degradation can be identified. Under emergency conditions limits are encompassed to which people may be exposed for short periods and survive without any permanently adverse effects. Environmental control systems are required to sustain a breathable atmosphere and to maintain the internal climate within the comfort zone. Logistic support is used for the provision of adequate food, disposal of waste products and transfer of personnel.[55]

6.1 ATMOSPHERE CONTROL

The critical elements of atmosphere control are oxygen supply, carbon dioxide removal and trace contaminant control with atmosphere analysis to ensure the safety of the contained environment. The manning levels of the structure and the duration of submergence will be the predominant factors in the selection and cost of suitable life support systems.

6.1.1 Oxygen Supply

Oxygen is present in the earth's atmosphere at about 21% by volume and has a partial pressure of 159 mm of Hg at a pressure of one standard atmosphere (760 mm of Hg). Variation in the range 18 - 24% (137 - 182 mm Hg) produces no significant physiological effects; partial pressures above 760 mm Hg and below 113 mm Hg have serious health effects. On average a man in a restricted environment consumes oxygen at a rate of 1 standard cubic foot per hour, which is equivalent to 0.9 kg (2 lb) of oxygen per man-day. The consumption will obviously be dependent on work load and dietary balance (Table 6.1.1).

TABLE 6.1.1 Oxygen consumption.

Oxygen Consumption	*Ideal*	*Normal Max.*	*Normal Min.*
kg/man-day	0.9	1.6	0.5
(lb/man-day)	(2.0)	(3.5)	(1.2)

Oxygen can be supplied to the underwater structure for atmosphere make-up using storage systems and transport vehicles, or the oxygen may be produced within the structure using oxygen generation equipment. The method used will be determined by the logistic support considerations, manning levels and submerged duration. Systems that may be considered are:

1. High Pressure (Gas) Oxygen Storage
2. Liquid (Cryogenic) Oxygen Storage
3. Electrolytic Oxygen Generation from Water
4. Chemical Oxygen Sources
   (a) Chlorates $2NaClO_3 \rightarrow 2NaCl + 3O_2$
   (b) Hydrogen Peroxide $H_2O_2 \rightarrow H_2 + O_2$
   (c) Potassium Superoxide $4KO_2 + 2CO_2 \rightarrow 2K_2CO_3 + 3O_2$

High pressure storage of oxygen is reliable and requires little power to operate the supply system, but is disadvantageous from the view point of weight and volume if large volumes of oxygen are required. Liquid oxygen offers substantial weight and volume savings and reduced operating pressures, but may suffer from heat leaks due to poor insulation. The re-supply of liquid oxygen by support vehicles will also present safety and operational problems. The use of electrolytic oxygen generators is a well proven technology but requires high electrical power levels for operation and hull penetrations for cooling and supply water and hydrogen dumping. Hydrogen peroxide is difficult to store and superoxide systems are still under development, although they offer the added bonus of converting CO_2 to O_2. Chlorate candles are a reliable and proven source of oxygen, but they may cause storage and disposal problems.

For the concept of a permanently manned underwater structure considered in this study, at present the electrolytic oxygen generation system would appear most appropriate with emergency back-up supplied by chlorate candles or externally stored high pressure oxygen bottles. The back-up systems could also be incorporated into the design of the safe haven areas to provide the emergency life support systems. Developments in the storage and logistic supply of liquid oxygen in the subsea environment should be investigated as this will generate an alternative choice of oxygen source with substantial power savings and reduced hull penetrations. Superoxides show considerable potential in offering the dual role of absorbing CO_2 and generating oxygen. The advantage to be gained if it becomes possible to match the reaction rate to the human respiratory exchange ratio such that O_2 can be generated and CO_2 absorbed in equilibrium is obvious and should also be investigated in detail.

6.1.2 Carbon Dioxide Control

The consumption of 0.9 kg (2 lb) of oxygen a day by a crew member will result in the production of 1.1 kg (2.4 lb) of CO_2 and water by the normal metabolic process. The respiratory exchange ratio averages out at about 0.85 but will be dependent on diet and activity level. The carbon dioxide control system should be capable of maintaining the CO_2 level below that which will impair mental and physical performance and ideally below the level in the earth's atmosphere of 0.17 mm Hg. Long-term exposures in enclosed environments will require levels of CO_2 to be below 0.5% (3.8 mm Hg), while higher levels can be tolerated on an intermittent basis (Table 6.1.2).

TABLE 6.1.2 Carbon dioxide production.

	Ideal	*Normal Max.*	*Normal Min.*
Carbon Dioxide Production kg/man-day	1.1	2.2	0.5
(lb/man-day)	(2.4)	(4.8)	(1.1)

The removal of the 1.1 kg (2.4 lb) of CO_2 produced per man-day from the atmosphere can be accomplished by several types of system. The systems used in spacecraft and submarines depend either on absorption or adsorption of the CO_2. Adsorption of CO_2 is usually a reversible process brought about by heating, exposure to vacuum, high air flow rates or a combination of these methods. Absorption systems involve chemical combination and because of the high energy required to reverse the process they are considered irreversible.

1. Metallic Absorbents
   $2XOH + CO_2 = X_2CO_3 + H_2O$ (Lithium Hydroxide, Soda Lime etc)
2. Molecular Sieves — (Regenerative) — Adsorption
3. Monoethanolamine Scrubbers — (Regenerative)
4. Absorption of CO_2 and Regeneration of O_2.

   (a) Sabatier Reaction $CO_2 + 4H_2 \xrightarrow[232°C]{Ni} CH_4 + 2H_2O$

   (b) Bosch Reaction $CO_2 + 2H_2 \xrightarrow[704°C]{Fe} 2H_2O + C$

   (a) and (b) these electrolyse H_2O to give O_2

   (c) Solid Electrolyte $2CO_2 \xrightarrow[538°C]{Elect.} 2CO + O_2$

   (d) Molten Electrolyte $LiCO_3 \xrightarrow[2200°C]{} LiO_2 + C + O_2$

Lithium hydroxide would appear to be the best metallic hydroxide absorber when cost, availability, CO_2 capacity and heat of absorption are considered and is currently widely in use. Volume requirements for large manning levels and long submergence times become restrictive and the material also exhibits some irritant qualities. Molecular sieve systems have a high affinity for CO_2, water vapour and other substances and preferentially absorb water so the air has to be dried before CO_2 is removed. Although its capacity to absorb CO_2 is lower than lithium hydroxide the molecular sieve is readily regenerated with heat or a vacuum. The monoethanolamine scrubber is also regenerative and currently used successfully in nuclear submarines and will absorb 12 lb CO_2 per hour at a flow rate of 250 cuft per minute. The power requirement, however, is high (11 kW) and the low iron monoethanolamine solution deteriorates with time. Both systems require the dumping of CO_2 driven off during regeneration. The generation of O_2 from CO_2 has obvious advantages; the technology, however, is still under development. The Bosch system would seem to hold the best promise as it can also be operated in the Sabatier mode through temperature control. Superoxide systems should also be investigated. In this application the practical experience of the use of the scrubber system would appear to make it most suitable, although recent development with molecular sieves should be evaluated. Emergency and back-up systems could be satisfied with lithium hydroxide canisters at present, although developments in superoxide systems should also be investigated. The permanent station concept may justify the cost and complexity of the Bosch/Sabatier systems and the state of the art of these systems should also be reviewed.[62]

6.1.3 Contaminant Control

The atmosphere of the sealed environment of an underwater structure will contain trace quantities of gaseous and particulate matter emanating from the crew and the materials within the enclosure. In deep ocean structures it may not be possible to vent the atmosphere intermittently to reduce the contaminant level so internal systems will be needed to control the level of contaminants. The type and toxicity of the contamination will determine the maximum permissible concentration acceptable over any given time period. These will be dependent on the constituent materials of construction of the structure and equipment, materials in use in structure and the crew and their life style. The most effective contamination control will therefore be the careful selection of construction materials and materials allowed into the environment (Table 6.1.3).[68]

TABLE 6.1.3 External control.

(a)	Control of construction materials.
(b)	Screening of stores to be taken onboard.
(c)	Reference laboratory analysis of samples of structure atmospheres.

Many of the contaminants will be difficult to detect and identify and the effect on crew performance will be dependent on their toxicological properties. The toxicity of many contaminants has been established for industry (40 hour week) but in many cases the effect of continuous exposure is not known. Submarine work has, however, established toxicity levels for many common contaminants for long term continuous exposures (90 days). The effects beyond these periods are still unknown.

The candidate methods for the control of contamination levels would be:

1. Absorption
2. Oxidization
3. Filtration
4. Electrostatic Precipitation

High boiling point organics and noxious odours will be absorbed by activated charcoal and molecular sieves can be used to absorb hydrogen sulphide, ammonia, ethane, acetylene, water and carbon dioxide. Carbon monoxide, methane and hydrogen need to be removed by oxidization using a catalytic burner; however, acidic gases may be produced as a by-product and these need to be taken up in a charcoal filter. Fibreglass filters can be used to remove particulate matter greater than 0.3 microns and aerosols less than

0.3 microns can be removed by electrostatic precipitators. It can be seen that no one method is suitable for all contaminants so a system incorporating all the methods will be required.

In this concept all contaminant control systems will also be required. Contaminants within the structure as a result of contruction materials, oil processing system, life-style etc. will need to be investigated in detail, to identify specific contaminants and to establish acceptable toxicity levels for each for short term occupancy by the crew (2 weeks). It will also probably be necessary to incorporate a temporary umbilical system to the surface for periodic venting of the structure to reduce contamination level build-up over long periods of submergence.

6.1.4 Atmosphere Analysis

The integrity of the atmosphere of a manned underwater structure will be highly dependent on the adequacy and accuracy of monitoring and analysis equipment. The measurement of levels of oxygen and contaminants will be required to maintain a breathable atmosphere (Table 6.1.4). The complexity of atmosphere analysis equipment that will deal efficiently with the task will be determined not only by the total man days of occupancy but also the total length of the occupancy.

TABLE 6.1.4 Internal control.

(a) Air treatment system producing oxygen and removing carbon dioxide and other gaseous and solid contaminants.
(b) Continuous monitoring of major constituents .
(c) Intermittent monitoring for minor contaminants.

Analysis and monitoring equipment generally exist to meet the requirements of manned underwater structures. However, structures remaining permanently submerged and unvented could pose several problems. A method will be needed to measure the potentially hazardous trace contaminants which may accumulate over long periods. The nature of some of these potential contaminants is such that their tolerance limits, method of detection and methods of control have at present not yet been resolved. If it is not possible to intermittently vent the atmosphere of the structure these potential difficulties may severely inhibit the structure's role. A wide range of atmosphere monitoring and analysis equipment will be required to control the various constituents of the contained atmosphere.

Oxygen content can be measured by simple battery powered and manually operated systems or by continuously operating line-power precision systems. The galvanic cell is battery operated and measures percentage volume of oxygen. The polarographic sensor, also battery powered, gives a continuous indication of the partial pressure of oxygen. The system which gives maximum reliability over long periods is a line-powered instrument which relies on the paramagnetic effect of oxygen in an electric field.

The simplest method of measuring CO_2 level requires no power and no calibration. A pre-prepared sample tube (Draeger Tube) using a 'Wet-Chemical' technique gives a colour indication of CO_2 level, however these tend to be inaccurate especially at low levels of CO_2. Infrared analyses operating on line-power are sensitive and stable, circulated air is tested for the absorption of an infrared source, the absorption being proportional to the CO_2 content of the air.

Complex analysing equipment operating on line-power has been developed for the nuclear submarine programme. The early models measured CO_2 by means of infrared absorption, oxygen by paramagnetic detection and hydrogen by thermal conductivity. Latest developments involve the use of gas chromatography to measure CO_2, CO, O_2, H_2 Freon 11 and Freon 12 and give greater sensitivity.[55]

The availability of line-power in our concept would suggest in the simplest form paramagnetic sensors would be used for oxygen measurement and infrared detectors for carbon dioxide control. Recent developments in gas chromatography systems should, however, be investigated in detail as these may offer an integrated atmosphere analysis system and the possibility of detection and measurement of trace contaminants. Battery powered portable instruments for the measurement of O_2, CO_2, H_2S and explosive limits for hydrocarbons will be needed for use in emergency situations and for monitoring areas prior to manned intervention if these areas are not serviced by permanent monitoring instruments.

6.1.5 Emergency Life Support Systems

In an enclosed environmental system any malfunction of the atmosphere control system or gross contamination of the atmosphere by leakage, smoke or volatile liquids will require the use of an emergency breathing system. In the event of a fire occurring the use of carbon tetrafluoride and carbon dioxide extinguishers will be in doubt because of the danger of contaminating the atmosphere and the limitations of the atmosphere purification systems.

Emergency procedures and methods of escape from the structure in the case of a catastrophic accident will need to be evaluated so that suitable emergency life support systems are availailable in these circumstances. Safe areas where the crew can await rescue will also need to contain their own life support systems and analysis equipment. The equipment should be independent of normal operating systems and easily operated, maintained and tested. All systems should be clearly labelled and include simple operating instructions. Basic facilities should be available in waiting areas for waste management and odour control. Warmth retaining clothing should also be provided with suitable prepackaged food and recreational equipment.

6.2 FOOD MANAGEMENT SYSTEMS

The nutritional requirements for men living and working within an underwater structure will be similar to the requirements for men performing like tasks in a comparable environment on the surface. The length of occupancy will not materially alter the dietary requirement, although the methods of preparing and packaging the food will be influenced by the degree of logistic support available. The dietary balance will, however, influence the total weight of food, the oxygen required to oxidize the food and the amount of carbon dioxide and water produced by the oxidization process. The food requirement will also be affected by the environmental temperature, the degree of physical activity engaged in and the physiological characteristics of the individuals. These factors and methods of food preparation will obviously have a secondary effect on life support parameters.[69]

The methods of packaging and storing the food will be influenced not only by the need to provide a normal diet in nutritional value, but also in appearance and taste. There are a number of basic candidate food types:

1. Dehydrated Foods
2. Freeze Dried Foods
3. Frozen Foods
4. Conventionally Packaged Foods (Tins, Dry-packaged etc.)
5. Fresh Foods

The choice of packaging will depend on such factors as palatability, keeping qualities, weight and volume and ease of preparation. A wide variety of food types will probably be required to provide a palatable diet with the bulk probably being frozen, freeze dried and conventional dry storage types. The use of fresh foods will be dependent on the frequency of logistic support operations. Facilities for food preparation with personnel for these tasks will be a basic requirement. The food requirements on a man-day basis are essentially fixed and therefore the weight, volume and storage requirements will affect overall system design directly.

6.3 WATER SUPPLY AND WASTE MANAGEMENT

Water will need to be provided for drinking, cooking, personal hygiene and laundry facilities and requirements will depend on the number of occupants, length of stay and special operational requirements of the structure. The water may be supplied from storage, desalination of seawater, condensation of humidity or reclamation of waste water. Waste management and water management are closely related and optimum water management should take into consideration the dumping or reclamation of waste water. For example, appreciable economy in water consumption may be affected by using shower and laundry water to flush toilets and atmosphere condensates to supplement the basic water supply.

Water supply systems that can be considered are:

1. Stored Water
2. Sea Water Distillation

3. Reverse Osmosis
4. Electrodialysis
5. Reclamation
6. Multifiltration

Stored water systems use minimum power and are simple in operation. The logistics of resupply may however cause difficulties, although flexible or other suitable tanks could be towed to the structure and the contents transferred to holding tanks within the structure. A number of distillation processes exist for the conversion of salt water to fresh water. Vapour distillation techniques have been well developed for shipboard use, they consume low power and add little heat to the environment. The system, however requires hull penetrations for seawater inlet and overboard pumping of the concentrated brine solution against ambient sea pressure. The reverse osmosis system uses a permeable membrane to separate fresh water from salt water using pressure, which could be derived from ambient sea pressure and is simple, economic and operates at ambient temperatures. Electrodialysis is at present a prohibitively expensive technique at salt concentrations of seawater. A degree of reclamation of water for secondary purposes could provide economies in water use and multifiltration using activated charcoal resin beds could be used to improve the quality of water from reclamation systems.

The use of stored water systems is simple and economic and preferred until weight, volume and support requirements become restrictive. Larger demands should be met with seawater distillation using vapour distillation techniques, providing hull penetrations and overboard dumping are acceptable. If not, waste water should be reclaimed by using combinations of multifiltration and vapour distillation techniques. Humidity condensation can be used to supplement the primary water supply. The reverse osmosis technique is attractive because of its low power consumption and normal operating temperatures.

Generally for large systems of long submergence, providing hull penetrations and overboard dumping are acceptable, reclamation cannot compete with seawater distillation or reverse osmosis. The power requirement to operate pumps for overboard dumping against ambient sea pressure will be significant and should be considered in the overall system design. It is not envisaged that full reclamation, including urine to potable water, will be required; some reclamation will assist in water management.

6.4 HEAT MANAGEMENT

The requirements for the heat management of a large manned underwater station are dependent upon several factors which include:

1. Ambient seawater temperature
2. Total hull surface area
3. Hull insulation
4. Internal heat generating equipment

The ambient temperature of the seawater environment will be a function of water depth and site location, hull surface area will depend on the geometrical configuration of the structure. The type and degree of hull insulation required will be determined by the material of construction of the structure and will be used to control heat transfer to the seawater environment and internal condensation. Heat will be generated within the structure by operating machinery, life support systems and other auxiliary equipment. Temperature and humidity for the manned environment should be maintained between 20 – 25°C with a relative humidity between 50 – 70%. Lower humidities should go with higher temperatures and vice versa. Condensation should be avoided because of the danger to equipment operation and the possibility of fungal growths. Thermoelectric heating and cooling will probably be the best method of temperature control when dealing with an array of heat generating equipment.

In the concept of a total subsea oil production system it is envisaged that power requirements may be of the order of 30 MW for operating equipment with additive heat loads from flows of hot oil. The major problem will therefore be the removal of thermal energy liberated within the structure. If active seawater cooling systems are required to shed the heat load, the biological and physical implications of the introduction of this heat to the surrounding seawater environment should be investigated in detail.

7

PHYSIOLOGICAL IMPLICATIONS OF ATMOSPHERE CONTROL

The composition of the atmosphere of the underwater structure must be such that the crew of the structure over a given time period do not suffer discomfort or loss of wellbeing during the exposure itself and neither deficit in health throughout the whole of their lives nor any life shortening effects. An optimum atmosphere will not be provided by simply maintaining the correct level of oxygen and removing the products of body metabolism such as carbon dioxide and carbon monoxide. The composition of the enclosed atmosphere will eventually reflect all materials involved in its construction, its contents, and all activities carried out within the structure. This section attempts to briefly assess acceptable levels of atmosphere constituents and the physiological implications of deviations from these levels.

7.1 OXYGEN LEVEL

Normal atmospheric air contains 79% nitrogen, 20.9% oxygen and 0.10% carbon dioxide and inert gasses together with water vapour. The partial pressure exerted by the oxygen in normal air is 159 mm of Hg. The human body is dependent on the partial pressure of oxygen, rather than the percentage of oxygen in the breathing atmosphere i.e. aviators need enriched mixtures of air/oxygen at reduced pressures and divers need smaller volumes of oxygen under higher pressures. Oxygen may be supplied by itself at pressure, or in a mixture of gases.

Oxygen that is stored in the body is in the blood, and the tissues that are most sensitive to oxygen supply are those of the visual and central nervous system. The brain, which only accounts for 2% of body weight, consumes 20% of the total body oxygen supply. If the partial pressure of oxygen falls below 132 mm of Hg symptoms of hypoxia set in and headaches, drowsiness, a general feeling of malaise, nausea and vomiting soon develop and, if the condition persists, neurological symptoms will occur causing a reduced capacity for mental work, auditory and visual disturbances and physical difficulties. Restoring the level of oxygen will usually alleviate all symptoms rapidly without the need for additional treatment, if the exposures are not prolonged.

At elevated levels of oxygen partial pressure, 1500 mm of Hg, haemoglobin in the body tends to become saturated and sufficient oxygen can be carried by the plasma portion of the blood to satisfy body needs. Little haemoglobin will therefore be reduced and as a result the carbon dioxide carrying capacity of the haemoglobin will be impaired and carbon dioxide will accumulate in the tissues, after several hours the signs and symptoms of acute oxygen poisoning will develop. Dilation of the pupils occurs, central vision is impaired and eventually blood pressure and pulse rate increase and convulsions rapidly ensue. Restoring oxygen levels to normal alleviates these conditions and prompt and complete recovery occurs within an hour or two.[72]

Prolonged exposures to partial pressures of oxygen between 140 mm and 160 mm Hg (18.4% to 21% at one atmosphere) have been found not to cause any alterations in physiological functions or the development of subjective symptoms, which can be attributed to the lack of oxygen. The experience of space work is shown in Table 7.1.1 which shows the drift from the pure oxygen environments (Apollo) to terrestrial like atmosphere systems in common with Russian designs.[65] The permissible range for oxygen, produced by electrolysis from seawater, is 137 mm Hg partial pressure and 22% by volume (167 mm) in Royal Naval nuclear submarines.

The lower limit is that below which a deterioration in night vision is expected to occur and the upper limit is set by fire

TABLE 7.1.1 Space craft atmosphere comparisons.

System *Parameter*	*LM*	*Soyuz*	*Salyut*	*Skylab*	*Shuttle*	*Spacelab*
Crew time	48 man-hr/day	48 man-hr/day	72 man-hr/day	71 man-hr/day	28 man-days	52 man-hr/day
Crew size	2 men	2 men	3 men	3 men	4–14 men	3 men
Duration	10 days		23 days	186 days	7–30 days	7–30 days
Atm. total press	248±10.3mm Hg	530–760mm Hg	760–960mm Hg	258±10mm Hg	760±10mm Hg	760±10mm Hg
O_2 partial press	248±10.3mm Hg	144–170mm Hg	160–280mm Hg	186±15.5mm Hg	160±5.2mm Hg	165±13mm Hg
N_2 partial press	0	Difference	Difference	Difference	Difference	595±13mm Hg
Volume	6.7 m^3		369 m^3	361 m^3	56.6 m^3	42.7 m^3
CO_2 partial press	7.6mm Hg Max		9mm Hg	5.5mm Hg Max	7.6mm Hg Max	5–7.6mm Hg

risks, particularly in hydrocarbon laden charcoal filters. In our concept these levels would appear appropriate, although it may be worth considering specific oxygen levels for different modules of the complex. Table 7.1.2 summarizes preferred and emergency oxygen and carbon dioxide levels.

TABLE 7.1.2 Normal and emergency oxygen and carbon dioxide levels.

	Partial pressure (mm Hg)				
	Ideal	*Normal Max.*	*Normal Min.*	*Emergency Max.*	*Emergency Min.*
Oxygen supply	159	167	134	760	113
Carbon dioxide control	0.17	3.8	0	34	—

7.2 CARBON DIOXIDE LEVELS

The terrestrial atmosphere contains less than 0.1% carbon dioxide, the concentration of carbon dioxide in alveolar air is approximately 5%, which is due to the gaseous diffusion of carbon dioxide from the pulmonary capillary bed. As the concentration of carbon dioxide in the inspired air is increased it becomes progressively unfavourable for the normal diffusion of carbon dioxide from the blood. Initially the body will compensate with an increased respiratory depth and rate until the increases in cardiac and respiratory rates cannot cope effectively. Carbon dioxide will accumulate in the blood and other body tissues. Under these conditions the acute effects of carbon dioxide exposure will become manifest. After several hours at 2% carbon dioxide, headaches appear and respiration increases, at 3% there is a marked increase in rate and depth of respiration, headaches become more severe and diffuse sweating occurs. At the 5% level mental depression occurs and at a concentration of 6% visual disturbances and tremors develop. Unconsciousness usually occurs at about the 10% level, with CO_2 acting as a narcotic at high concentrations (Table 7.2).[73]

TABLE 7.2 Effects of acute exposure to carbon dioxide.

Level (%)	*Symptoms*	*Remarks*
1–2	Slight increase in depth of respiration. Headache. Fatigue (chronic after several hours).	
3	Severe Headache. Diffuse sweating.	Acceptable limit of alertness
4	Flushing of face. Palpitations.	
5	Mental Depression.	
6	Hard work impossible. Visual disturbance.	
8	Tremors. Convulsions.	
10	Unconsciousness.	

Acute effects are characterized by exposure to high levels of carbon dioxide for short time periods. Chronic toxic effects can produce widespread physiological alterations after exposure for long periods to elevated levels of carbon dioxide, well below the levels at which acute effects are noticeable. The toxic effects therefore have to be investigated in terms of the life-time health of the crew, regarding early development of degenerative diseases or actual life-shortening. Present evidence suggests that even at levels of carbon dioxide of 1% (7.6 mm Hg) changes in mineral metabolism, respiratory physiology and acid–base balance occur, which take many weeks to normalize after removal from the enclosed atmosphere. Recent experimental work[74] with carbon dioxide levels as low as 0.5% (3.8 mm Hg) has shown that possibly medically significant changes are demonstrable at this lower level.

The work is also complicated by the concomitant presence of abnormal levels of carbon monoxide and other atmosphere contaminants and the lack of sunlight which reduces Vitamin D levels.

The present level of carbon dioxide established for submarine and space operations is 1% (7.6 mm Hg). On the basis of recent work it would appear that a maximum level of 0.5% (3.8 mm Hg) would be advisable for extended periods of enclosure. Ideally a terrestrial level of carbon dioxide would be required; however, life support equipment may limit the achievable rate of carbon dioxide removal. The physiological implications of crew rotation every two weeks should be investigated to determine the significance of a two week exposure and two week recovery sequence to the allowable maximum permissible concentration (M.P.C.) of carbon dioxide. It may be possible to relax the M.P.C. under these conditions.

7.3 CARBON MONOXIDE LEVELS

Carbon monoxide is not normally present in the atmosphere, but is produced whenever any carbonaceous material is oxidized incompletely and is the most important gaseous poison as it accounts for more deaths than all other toxic gases combined. The toxicity is due mainly to its ability to displace oxygen in the blood as it has two hundred times more affinity for haemoglobin than oxygen. The mere presence of small amounts of carbon monoxide in the atmosphere will markedly reduce the capacity of blood to absorb and transport oxygen. The percentage of haemoglobin that is reduced to carboxy haemoglobin (COHgb) during exposures to known concentrations of carbon monoxide over various time periods can be determined (see Table 7.3). The level of carboxy haemoglobin in the blood has been correlated with the onset of subjective symptoms and physiological conditions.[72]

The major source of carbon monoxide in nuclear submarines (75 - 90%) is caused by tobacco smoking, however, the carbon monoxide level will increase even if smoking is prohibited. Carbon monoxide will be evolved from the oxidization of oils and lubricants especially on hot components, from the aging of oil based

TABLE 7.3 Percentage of carboxyhaemoglobin (COHgb) and symptoms following an 8-hour exposure to carbon monoxide.

CO in air (ppm)	*COHgb (%)*	*Symptoms*
50	0–10	No symptoms.
100	10–20	Tightness across the forehead, possibly slight headache, dilation of the cutaneous blood vessels.
200	20–30	Headache and throbbing in temples.
300	30–40	Severe headache, weakness, dizziness, dimness of vision, nausea, vomiting, and collapse.
500	40–50	Same as above, a greater possibility of collapse, syncope and increased pulse and respiratory rates.
750	50–60	Syncope, increased respiratory and pulse rates, coma, intermittent convulsions and Cheyne-Stokes respiration.
1000	60–70	Coma, intermittent convulsions, depressed heart action and respiratory rate and possibly death.
1500	70–80	Weak pulse, slow respiration, respiratory failure and death within a few hours.
2000	80–90	Death in less than an hour.
4000	90+	Death within a few minutes.

paints and from the endogenous production in the crew and other activities. It will therefore be necessary to remove carbon monoxide even if smoking is banned.

The maximum permitted level of carbon monoxide must be selected, so that the level of induced carboxy haemoglobin in association with other adverse environmental factors present in the enclosed environment does not cause acute illness or chronic disease or play a part in the development of degenerative illness such as cardiovascular disease. The crew should also be able to do normal tasks without any decrement in performance. If it is assumed that smokers have compensated physiologically and mentally for their self-induced COHgb burdens, the maximum permissible concentration should be the highest level at which COHgb in non-smokers has been shown not to produce a performance decrement or an increased risk of cardiovascular disease.

Vigilance or general sensitivity of the central nervous system to incoming sensory information is generally agreed to be affected by COHgb levels well below 5% (30 p.p.m.) and is the factor probably most applicable to our situation. On the basis of past studies COHgb level of 2.5% corresponding to a carbon monoxide level of 15 p.p.m. would appear to be recommended. There is probably no above-endogenous level of COHgb at which there is no risk of cardiovascular disease. Past work shows a substantial increase in disease risk associated with continuous COHgb loads above 3%, and allowing for the presence of other health hazards a COHgb burden of 2.5% is again recommended as the maximum concentration.[75]

The present carbon monoxide maximum permissible concentration for a 90-day exposure for United States and Royal Naval nuclear submarines is 25 p.p.m. although a reduction to 15 p.p.m. has been recommended and they are at present operationally achieving levels between 5 - 15 p.p.m. for the majority of the patrol. This figure agrees with proposals for spacecraft for 90 and 1000 day missions of 15 p.p.m. This figure should easily be achieved within our concept using current equipment and the problem may be eased by restricting smoking, although this may have an adverse effect on crew performance or morale.

7.4 OTHER ATMOSPHERE COMPONENTS

The enclosed atmosphere of an underwater structure will eventually reflect all the materials involved in its construction, its contents and all the activities carried out within it. Several hundred different gases are believed to be present in closed-down nuclear submarine atmospheres and many atmosphere contaminants are regularly found in submarines and space vehicles. The complex mixture of gaseous contaminants is generated by many processes; smoking, aging of paints, resins, adhesives, oils, solvents etc. and life support machinery (Table 7.4).[70]

TABLE 7.4 Major atmosphere contaminants and their sources (Submarines).

Acetylene	Welding or burning equipment
Ammonia	Carbon dioxide scrubbers
Arsine	Battery gassing
Benzene	Paints, solvents, oils
Carbon dioxide	Respiration, smoking
Carbon monoxide	Smoking, lagging, fires, oils, paints
Chlorine	Oxygen candles
Freon	Refrigerant and air conditioning systems
Hydrocarbons	Paints, solvents, oils, cooking
Hydrogen	Battery gassing, electrolysers
Hydrogen chloride	Freon decomposition
Hydrogen fluoride	Freon decomposition
Mercury	Batteries, lighting, thermometers
Methane	Sanitary tanks
Methanol	Paints, cleaning materials
Methyl chloroform	Solvents, adhesives
Monoethanolamine	Carbon dioxide scrubbers
Nitrous gases	Carbon monoxide burner effluent, precipitators
Stibine	Battery gassing
Sulphur dioxide	Sanitary tank, carbon monoxide burner
Toluene	Paints, solvents, oils
Triaryl phosphates	Compressors, oils
Xylene	Paints, solvents, oils

7.4.1 Hydrocarbons

It is not possible to differentiate between the many hydrocarbons present in the atmosphere using shipboard equipment. Techniques are available to measure total hydrocarbons and aromatic hydrocarbons as groups. Proper analysis can only be achieved by using laboratory methods of mass spectrometry, infrared spectrometry or gas chromatography. Analysis has shown that 55% of total hydrocarbons in a stabilized atmosphere are aromatic, but wide variations occur when fried foods are prepared, producing aldehydes and unsaturated aliphate compounds. The detection and quantitative measurement of this group of hydrocarbons is difficult at present.

Aromatic hydrocarbons are chiefly derived from the destructive distillation of coal and fractionalization of petroleum. They include benzene, toluene and xylenes which are used in paints, enamels, solvents etc. The aromatic compounds gain access into the atmosphere at a slow logarithmic rate after the application of the material containing them. Acute poisoning due to this contamination is unlikely, but chronic effects due to long exposures to small amounts of these substances could present a hazard. Chronic intoxication would show symptoms of suppressed bone marrow activity, with a resultant anaemia and loss of blood clotting ability and reduction in the number of white blood cells.

The problem of aromatic hydrocarbon contamination can be alleviated by the use of water-based rather than oil-based paints and lacquers and the restriction of the use and storage of paints, solvents and cements. Constant control of materials introduced in the enclosed environment by crew indoctrination and comprehensive materials guides would appear to be the most efficient way to control these atmosphere contaminants from previous experience. They are not readily removed by atmosphere control equipment.

7.4.2 Freon

Freon is the name given to a group of chlorinated or fluorinated hydrocarbons and they are used as refrigerants and aerosol propellants. They are non-inflammable and exhibit low toxicity, the main danger lies in their chemical properties. Freons are decomposed by heat and when incompletely oxidized produce chlorine, fluorine, hydrogen chloride, hydrogen fluoride and possibly phosgene. The maximum permissible concentrations for Freons are therefore set to maintain the degradation products within their threshold values (F_{11} = 5 p.p.m., F_{12} = 500 p.p.m.). The problem can be alleviated by correcting leaks in operating equipment and restricting spillage during repair and maintenance procedures. The halogen and halogenic acids produced will be readily absorbed in the carbon dioxide scrubber.

7.4.3 Aerosols

The concentration of aerosols found in submarine environments is greater than that in rural communities but less than that in industrially polluted areas. However, the major portion of the aerosols present are organic and therefore many times more toxic than industrial air pollution. Aerosols cause difficulties in the operation of electrical and electronic equipment as well as being a health hazard. Three quarters of aerosols are produced by cigarette smoking and the remainder by the nebulization of lubricating oils and from cooking fumes. The catalytic combustion in the carbon monoxide burners of the aerosols combined with the filtering action of the electrostatic precipitators reduces the concentrations to acceptable levels.

7.4.4 Inorganics

Toxic inorganic and elemental substances will be present in the atmosphere. They are generally more easily controlled as more is known about them and they are readily identified (Table 7.4.4).[70]

Electrical arcing around motor armatures produces ozone and oxides of nitrogen. Nitrous oxide is relatively non-toxic, but

TABLE 7.4.4 Comparative M.P.C.'s for some submarine contaminants.

Substance	*M.P.C. 90 (mg/m^3)*
Ammonia	18.00
Carbon monoxide	27.50
Chlorine	0.30
Freon 12	2500.00
Hydrogen chloride	1.50
Hydrogen fluoride	0.10
Mercury	0.01
Methanol	13.00
Nitrogen dioxide	1.00
Ozone	0.04

nitrogen dioxide and ozone are highly irritant to all mucosal membranes. Fortunately ozone is highly reactive and is soon converted to molecular oxygen, nitrogen dioxide is readily absorbed in the alkaline solution of the carbon dioxide scrubber.

The bacterial action within the submarine's sanitary tanks produces sulphur dioxide and hydrogen sulphide. These gases find their way into the submarine atmosphere, when the tanks are vented inboard to equalize pressures after being emptied. The low concentrations in which they appear are easily removed in the carbon dioxide scrubbers.

7.5 MICROBIOLOGY

The enclosed environment of an underwater station is a closed community. The crew live in close quarters and a ready exchange of body flora will occur. The climate is relatively constant and the atmosphere will contain many chemical contaminants, both of which may affect the survival of organisms. There is an absence of the normal bacterial action of ultraviolet light although the presence of air purification equipment (catalytic burner) may compensate for this. The crew may be intermittently exposed to airborne concentrations of micro-organisms at a higher level and for longer periods than elsewhere. The potential for the spread of infection amongst the crew is high. Experience from the nuclear submarine programme shows that there is little evidence of an increase in either respiratory or intestinal infections and that reported sickness tends to fall-off during the patrol. Investigations have shown, however, that a shift in flora does occur, which may be a function of the aerosols generated or the preferential survival of particular species under altered ambient conditions. Airborne bacterial levels do, however increase significantly when sewage tanks are vented into the submarine atmosphere as would be expected.[60]

Microbiological control can be enhanced by medical inspection of on-going crew prior to structure entry to preclude the transfer of pathogenic organisms into the environment and the use of bacterial filters and/or activated charcoal to clean lavatory exhausts and galley exhausts. Fungus growth, which is only a low health risk, can be restricted by maintaining suitably high temperatures and ventilation rates, and the use of fungicide paints.

7.6 CLIMATE OF ATMOSPHERE

The control of the climate of the enclosed environment is not only important in its own right for crew comfort, but is also a factor that may alter or modify the effects of other physiological stresses. Air conditioning is particularly vital for the high heat loads envisaged and for the provision of suitable ventilation rates to prevent 'pocketing' of local contamination levels and to reduce fungal growths. Temperature and humidity should be maintained automatically between 20 - 25°C and 50 - 70%, respectively, and also be controllable locally within these ranges in selected areas for improved crew comfort.

7.7 RECREATION AND EXERCISE

Recreation is a significant part of man's living pattern and it is accepted as commonplace because of the ease in which it can be engaged. It is that time when one is not working, eating or sleeping. Opportunities for recreation will need to be planned into the system. Exercise is physiologically important for muscle tone, cardiovascular system and the optimizing of calcium metabolism in the bones. A suitable and acceptable exercise programme will need to be developed for crew members.

7.8 SLEEP, WORK/REST CYCLES

The design of suitable work-scheduling procedures will be important for the operational efficiency of the crew. The disruption of sleep, work/rest cycles and circadian rhythms will have a significant effect on performance. The task performance will also be a function of the degree of concentration required and the length of the duty cycle. Generally, performance is improved by alternate work and rest periods.

8
HUMAN FACTORS

The reliable and efficient performance of tasks within the environment by the crew will depend on their physical, physiological and psychological well-being and the system characteristics which affect that human performance. The requirements of life support and health maintenance have been treated earlier. Features of habitability and system design, as they relate to crew performance will now be investigated.

8.1 HABITABILITY

Basic engineering design considerations can have an overall effect on performance and the acceptance of the system by the crew. These factors should be optimized to effect a general feeling of well-being amongst the crew, they will be highly dependent on the tasks of the individual or group. Some of the basic factors that should be considered include:

1. Anthropometric factors
2. Noise and vibration
3. Illumination and colour
4. Psychological acceptance
5. Medical/Emergency Procedures

8.1.1 Anthropometric Factors

The crew will be composed of people of varying shapes and sizes. Facilities should be designed to take this into account so as to reduce the physical discomfort which may be accentuated over long periods of submergence and affect crew health and performance. Anthropometric data should be incorporated into the design of accommodation, services, work areas and emergency procedures.

8.1.2 Noise and Vibration

Noise: Operating equipment within the structure will generate acoustic noise. The human ear normally perceives sound between 20 Hz – 20 kHz and within this range it is important to keep noise levels within the limits for adequate performance. The subjective perception of sound depends on the frequency and intensity of the sound. Safe exposure levels will be determined by whether it is a continuous or an intermittent noise source. Noise does not necessarily disrupt performance but can distract persons receiving auditory or visual information.

Sound intensities between 30 – 50 dB are not disturbing nor do they interfere with normal speech communications, although special areas such as sleeping quarters should have reduced sound levels (< 30 dB) to assist crew comfort. Between 50 – 70 dB communication is difficult and exposures of 88 – 90 dB for a forty hour working week for several years will cause permanent hearing loss. The pain threshold is reached at about 130 dB and is only acceptable in emergency situations. Loud noises of short duration should also be avoided to reduce shock responses in the crew due to the operating environment.

Vibration: Mechanical vibrations in underwater structures could be a significant problem, producing severe performance degradations particularly in the frequency range below 100 Hz. Human tolerance levels to vibration are highly frequency dependent and can be annoying and damaging. Below 3 Hz the body behaves as a single mass without internal resonances. The abdominal thoracic contents have a natural frequency of 4 – 6 Hz and the human body has resonances at 5 and 12 Hz along the spinal axis. The frequency region between 2 – 6 Hz represents the natural frequency of the body and vibrations of this frequency can cause traumatic injury and should therefore be eliminated.

Low frequency vibrations will be a function of structural arrangement and mass, with localized high frequency vibrations being generated by operating equipment. Machinery generated vibrations can be reduced by the use of resilient mountings and other damping techniques to improve crew performance and comfort.

8.1.3 Illumination and Colour

Illumination will not cause any profound physiological changes, but will be related to performance especially in terms of visual efficiency. Good illumination is required for the effective performance of most work and attention should be given to many factors including:

1. Uniformity of illumination.
2. Suitable brightness contrast between work area and background.
3. A level of illumination suitable for the most difficult aspect of the job.
4. Lack of glare from work surface or light source.
5. Length of time on the job and accuracy required in the task performance.
6. Possible variations in operating conditions which may affect lighting conditions.

Glare is one of the more serious problems and is usually caused by brightly polished surfaces or highly reflective finishes. The use of a diffuse light source or matt surfaces can alleviate the problem. Contrast is also important, and suitable degrees of contrast between work area and background should be arranged.

Colour composition will also affect illumination as the amount of light reflected from a surface will be dependent in part on the colour. Several factors of psychological significance are associated with colour. Surrounds that are coloured red, orange and yellow are considered by many people to be stimulating and warm, whereas blue, green and violet are considered restful and cool. Pale colours are generally considered cooler than more saturated colours and bright colours are judged to be physically closer than is actually the case. A discordant colour scheme is dissatisfying or annoying to many people and may contribute to restlessness and loss of efficiency. Generally, restful colours should be recommended for large items or surfaces within the structure (see Table 8.1).

TABLE 8.1 Human factor requirements.

Item	*Ideal*	*Normal Max.*	*Normal Min.*	*Emergency Max.*	*Emergency Min.*
Illumination (foot candle)	40	50	20	100	5
Acoustic Noise (dB)	50	88	30	95	
Vibration (1–100 cps g's)	0	10^{-2} g		0.4g	
Living Area (ft^2 /man)					
Berthing	100	150	50		
Messing	15	20	9		

8.1.4 Psychological Acceptance

Psychological acceptance problems can be defined as the frustratin of human need and is a function of man's motivational system. A highly motivated person may compensate to a considerable extent for poorly designed equipment, to maintain system operation. A person, however, who is dissatisfied with a machine's role due to status, survival fears or simply a desire to perform the role manually may not properly use equipment, which has been designed to fit all other criteria. It is important to appreciate when designing the system that man does form such opinions as his attitude will have a maximum impact on system performance.

8.1.5 Emergency/Safety/Medical Procedures

The crew will need to be assured that suitable facilities are available and that operational arrangements have been made to cope with emergency situations and rescue from the structure, otherwise morale will be adversely affected. Fire control equipment, emergency breathing apparatus and medical equipment should be available within the structure for use in accident situations. Rescue vehicles capable of operating at the installation depth of the structure should be on stand-by, and the structure should incorporate buoyant transfer capsules to allow the escape of the crew from the structure in the case of catastrophic accident. Special arrangements may be necessary for the transfer of injured personnel.

8.2 SYSTEM CHARACTERISTICS AND HUMAN PERFORMANCE

The satisfaction of the physiological and habitability requirements in the context of human ecology in an enclosed underwater structure will not alone guarantee efficient system operation. Inherent in the concept of a manned underwater structure is the fact that man is there to do something. Additional stresses will therefore be imposed on the crew during the performance of their tasks. Operational requirements, equipment usage and crew composition will introduce interfaces that involve new problem areas and stresses. The following broad treatment identifies some areas that may be of concern within our concept.

1. Operational Requirements.
2. Structural Design and Equipment Factors.
3. Skill Demands.
4. Personnel and Training.

8.2.1 Operational Requirements

The crew while on-station have to achieve performance objectives in order to satisfy the work schedule. The successful completion of these tasks and the ease with which this is achieved will be determined by overall system reliability and competitive agents restricting the attainment of these goals. The absence of unscheduled work loads and effective teamwork will minimize system stresses.

The mere occupation of the underwater structure will produce stresses due to the ever present fear of ambient pressure, fire or failure of life support systems etc. and these stresses will increase with the duration of submergence. Every effort should be taken to assure crew members of the system's integrity and their capability to survive and escape in an emergency situation. Confidence should also be generated in the safety of transport systems between the structure and the surface.

8.2.2 Structural Design and Equipment Factors

The structural configuration of the underwater structure will be determined by the complex integration of many requirements. The major restriction will be the provision and location of watertight bulkheads and safe areas for receiving rescue facilities. The crew should, however, be able to move easily about the structure and have rapid access to emergency equipment. Adequate access should be provided for preventive and corrective maintenance of equipment. Integrated areas should be allotted for work, cooking, sleeping and waste management etc. and individual crew members should be allowed a personalized space for their privacy requirement.

Controls and displays for equipment operation will require intensive analysis of the requirements and constraints of the system before they are designed. Reliable human engineering data already available should be incorporated into the analysis to produce an efficiently balanced overall system that minimizes man-machine interface stresses. The degree of automation incorporated will need to produce a compromise between boredom and an acceptable level of vigilance.

The majority of maintenance of the structure will have to be done *in situ*. Further restrictions will be imposed by the need to transport equipment to the site and the size limitations of hatch and water-tight bulkhead openings. A design incorporating no more than 2 or 3 components in simple redundancy and/or preventive maintenance should be considered. The maintainability of individual systems should also be a function of their criticallity, i.e. life support, high maintainability. Experience has also shown that the inclusion of a special 'habitat engineer' within the crew, whose only responsibilities are maintenance, improves confidence among other crew members and reduces stress. A good maintenance program is critical to system operation and crew morale.

8.2.3 Skill Demands

The skill requirements of a crew in an underwater structure can conveniently be broken down into three areas, psychomotor, perceptual and cognitive. It is not envisaged that any extraordinary psychomotor skills will be required within the structure, although the operation of logistic support vehicles will demand special skills. The operation and control of sophisticated instrumentation systems will require enhanced levels of perceptual abilities and will warrant detailed investigation. These systems will also introduce a paradox in that some cognitive skills will be replaced by machinery; the reduced crews, however, will have increased responsibilities over a larger number of performance parameters and systems. The overall system will therefore need to be designed so that the crew are continually aware of the system state, in case they need to override the system with a minimum of delay. The role of man, the allocation of various functions and the degree of automation must be researched for each specific application to find an optimum design to meet operational requirements.

8.2.4 Personnel and Training

The size and composition of the crew will be determined by operational requirements, engineering systems and psychological factors. The crew size in our concept will be principally determined by the system work load. It appears important to allot an engineer specifically for maintenance tasks within the structure, also activities associated with logistic support and food preparation should form part of the crew complement to reduce stresses on system operating crewmen.

The crew should be selected on the basis of physical and mental health and psychophysiological parameters compatible with operational requirements, preferably volunteers with experience of the submarine environment. Crews should be assessed before and during training using standard procedures and trained in groups that will eventually form the operational teams. One or two team members should carry-over to the succeeding team to retain operational continuity. Initially, training should be carried out on a land-based system which should include simulation equipment duplicating the operating system.[55,62,64,76–80]

9

PSYCHOLOGICAL ASPECTS

The biological adaptation of the crew to an enclosed underwater environment will be a complex function of the nature and severity of imposed environmental stresses and the adaptive capacity of the persons involved. The size of crew and duration of exposure will be critical to the adaptation process. Additional psychological stresses will be imposed if habitability and human factor requirements of the individuals are not satisfied. The characteristics of individual crew members, the group dynamics within the crew and leadership qualities will affect group harmony and morale.

The establishment of normal behavioural patterns is very difficult; however, the deterioration of psychosocial stability will usually manifest itself as anxiety, irritation, loneliness and in extreme cases as irrational behaviour. The alleviation of such symptoms must be achieved by choosing an optimum crew size for the submerged duration and careful selection of crew members. These criteria are examined briefly in association with general environmental considerations for producing a suitable psychological climate.

9.1 CREW SIZE AND DURATION OF SUBMERGENCE

These factors will have a crucial effect on the psychosocial atmosphere of the structure. In general, larger crew sizes (> 25 men) are preferred for longer duration missions (> 2 weeks). Submarine crew selection techniques should be suitable for the choice of crew sizes between 25 and 250. Smaller crew sizes especially on long duration missions should be chosen using more rigorous selection techniques. Crew size should be optimized to mission duration and preferably above a minimum number for suitable psychosocial relations.

9.2 CREW SELECTION/TRAINING

The stringencies of community living will be magnified in an enclosed underwater structure and the crew will also be mutually dependent on one another. They will need to adapt to living and working within confined quarters and have the ability to co-exist with other crew members under these conditions. The cohesion and good relations within the crew will depend on the personality characteristics of the crew members as well as the physical environment.

The crew should preferably be composed of volunteers, experienced in the underwater field and have a good record of past competence. Comprehensive medical and psychiatric screening procedures should be used in association with assessments of motivational reasons for volunteering. Positive motivation traits should be considered as well as negative traits. The crew member should also be tested for aptitude (maths, psychomotor etc.). Extrovert and introvert types should be equally acceptable as long as they do not annoy or irritate.

Periods of training should be used to provide an opportunity for the observation of crew members as a group, to acquaint the members with each other and to familiarize the group with the operating system. The training period will also foster group unity and develop feelings of security and therefore enhance successful system operation.

9.3 THE CREW AS A GROUP

The behaviour, compatibility and effectiveness of the crew will be directly dependent on that unique combination of individuals that constitutes the crew. The mere presence of the group will affect the crew member's performance and the behaviour of the group will affect his attitude. Differences in objectives among crew members and disharmony between members will result in failure in the achievement of goals.

Social, educational and intelligence levels will relate to the ability of the crew members to communicate with one another and informal social skills will have to be invoked for successful crew behaviour. The crew leaders' personalities will significantly affect the crew climate that will develop. Activities as a group can, however encourage crew members with complementary skills to assist each other in attaining goals. The daily contact between crew members of various backgrounds and outlooks can also provide mutual stimulation to the overall good of crew morale.

9.4 ENVIRONMENTAL

There are many factors that will promote a good psychological structure for the maintenance of the crew, many of these have been dealt with in the treatment of habitability and human factors. The conditions within the structure should resemble as far as possible those of an equivalent land-based system in terms of operating environment and activity schedule. Interest and motivation will need to be continually stimulated to counteract boredom and monotony. The work should be meaningful and as variable as possible. Monotonous tasks should incorporate feedback to stimulate vigilance and some work should be kept complex for interest sake, the aim being to reduce symptoms of anxiety, indifference and regressive behaviour.

Communications by voice or closed circuit television to the surface support vessels or a land base could alleviate feelings of social isolation, although consideration should be given to the censorship of incoming information. Such information may induce stresses in a crew member, and because of his position he may not be able to take any action.

Structures should be large enough to offer sufficient space for each crew member to satisfy his need. Privacy areas should also be available for each crew member to allow him to withdraw from the rest of the crew and alleviate feelings of irritability and hostility.

Psychological stresses will be induced through fears of danger in his operating environment and the use of special equipment, i.e. breathing masks etc. These should be minimized with proper training, education and confident leadership.

In our concept, it is envisaged that the crew will consist of 50 men submerged for a two week period. This combination of crew number and submerged duration is most unlikely to produce any overt psychological disturbances. Submariner selection criteria would appear to be suitable for our application. The use of medical, psychiatric, aptitude and motivation assessment on volunteer crew will enhance crew compatibility and minimize psychological problems. Training should be carried out in teams for the observation of group behaviour and to foster group unity.[52,55,62,76,77,80,83–86]

10

SUBMERSIBLES

The logistic support of a manned underwater structure will involve the transport of personnel and materials between the subsea complex and a surface support facility or a land base. The transfer of 'on-going' crew will probably be complemented by the removal of waste materials and the evacuation of the 'off-going' crew by the delivery of replacement materials. The degree of access to the structure will be determined by the docking method adopted. The transport system may also be required to rescue personnel from the structure in emergency situations.

A submersible vehicle suitable for the logistic support of an underwater structure must be capable of safe operation at the water depth of the complex and have sufficient payload capacity to enable it to perform the task requirements. The submersible must also be capable of docking with the structure and performing one atmosphere transfers of men and materials between the submersible and the structure.

The following analysis attempts to assess present submersible capabilities in relation to the probable requirements for logistic support of a manned underwater structure and predict possible concepts of submersible operation. The submersible systems are generally considered purely as a transportation system for men and materials.

10.1 SURFACE SUPPORT SYSTEMS

At present submersible operations are carried out in association with specialized surface support vessels. The submersible is launched from and retrieved to the vessel using many different methods of lifting technique of varying sophistication. There are many limitations to these techniques, the most critical being adverse weather conditions and operations are generally limited to sea state 5 – 7. As the size and weight of the submersible increases the operational problems become compounded. The major cost element of submersible operations in this mode will be determined by the costs of operating the support vessel, rather than the submersible itself.

A requirement will exist to transfer significant volumes of materials to the structure, which will require frequent submersible operations. It is most unlikely that present methods of surface support could cope adequately with the larger submersibles required for these tasks consistently, especially in adverse environments. It may prove necessary to consider the possibility of operating submersibles from intermediate subsea positions which are connected back to the surface or with an autonomous submarine operating directly from a land base or the surface.

10.2 DOCKING TO THE STRUCTURE

The present experience of dry hatch mating with undersea systems is generally restricted to hatch mating between submersibles/ bells and submarines or subsea completion systems. The submersible is normally fitted with a special mating skirt, whose diameter matches the access hatch of the underwater structure. The water entrapped in the skirt at ambient pressure on mating is pumped out and the air pressure adjusted to one atmosphere. Once the pressure in the skirt, submersible and structure are equalized hatch covers can be opened and personnel and equipment transferred. The seal is maintained and submersible locked in position by the strong force of ambient sea pressure acting on the skirt.

This technique is now standard practice on an intermittent basis; however, the success of the operation can be complicated by many factors. Interfacing the two hatches is sensitive to submersible and structure hatch orientation and marine growth or entrapped cables can disturb the sealing surfaces. The submersible is required to perform precision manoeuvring to interface the two sealing faces and there is always the danger that the submersible will impact the structure due to bad navigation or control error.

Logistically the most critical aspect of this docking technique is that equipment and materials widths are restricted to the diameter of the hatch and hatch arrangements may also limit longitudinal dimensions. This method of access would be suitable for the rescue of entrapped personnel in an emergency, but it exhibits many limitations for the day-to-day transport of personnel and materials. Improvements in operational procedures and mating techniques are required to achieve reliable and frequent interfacing over continuous periods of operation. An alternative method of docking could be the incorporation of a sea lock in the complex to receive submersibles or autonomous submarines.

10.3 SUBMERSIBLES AND BELLS/CAPSULES

The advent of North Sea oil and gas has given a boost to the construction of manned submersibles. The major customer for submersibles today is industry and the major user of industrially owned machines is the oil and gas offshore industry.[89] Diver lockout and dry transfer are increasing capabilities in industrial submersibles and generally they are larger and more powerful than before. The pressure hulls for depths greater than 600 m are normally spherical as this is the most efficient shape for dealing with the ambient pressures at these depths, although this reduces effective volume for personnel and equipment. Diving bells have also become more sophisticated, incorporating propulsion systems that allow the bell to manoeuvre at the end of its tether.

In contrast to past experience new submersibles and bells are being built to the specification of the purchaser and reflect his idea of the present and future capabilities required for subsea work. Submersibles are therefore evolving in a specific way to match requirements, so a totally new requirement such as the support of a manned underwater structure will probably involve a new design concept of a submersible to match its task.

The following table indicates the major characteristics of some advanced submersibles presently in operation or in a state of

TABLE 10.3 Submersibles with dry transfer.

Name	*Builder*	*Operator*	*Depth (m)*	*Weight (kg)*	**Payload (kg)*
Mermaid VI	Bruker Meerestechnik Germany	A Foreign Navy	600	17000	1000
L5	Vickers-Slingsby U.K.	British Oceanics	475	—	180–500
Taurus	International Hydro-dynamics Canada	British Oceanics	610	24000	800–1800
PC-16	Perry Oceanics U.S.A.	British Oceanics	915	15000	270
Deep Sea Rescue Vehicle (D.S.R.V.)	Lockheed Missile & Space	US Navy	1524	34000	1950
U.R.F.	Kockums, Sweden	Swedish Navy	460	49000	3000

*Payloads will vary as a function of incorporated operational equipment.

advanced design that offer the capability for dry transfer to an underwater structure (Table 10.3 and Fig. 10.3).[90,91]

The data shows the correlation between vehicle weight and

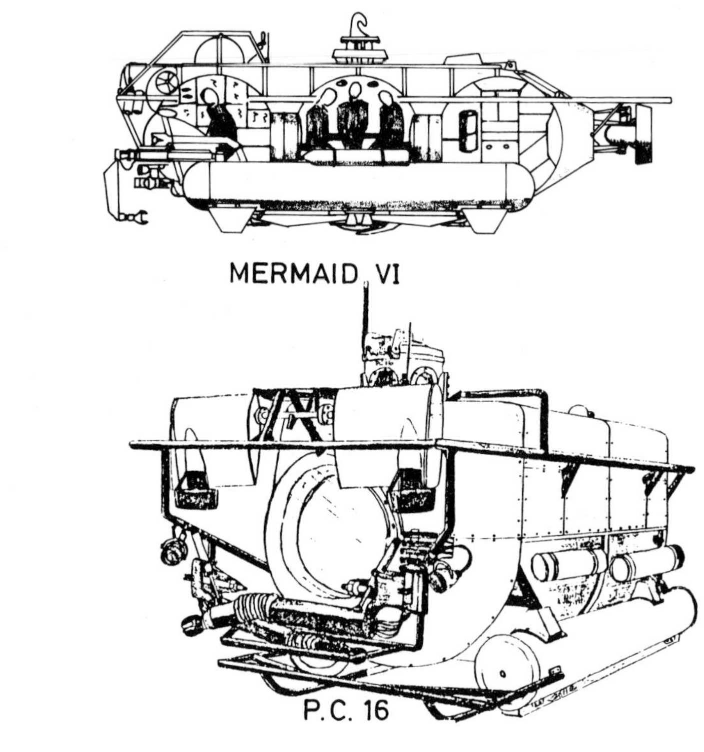

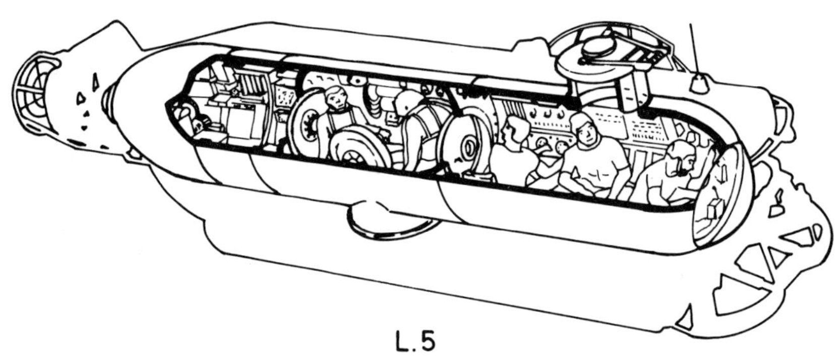

Figure 10.3 Submersibles.

payload capacity. Assuming a maximum crew size for the underwater structure of 50 men of a minimum average weight of 76 kg (12 stone) the on-going and off-going crew will represent a payload of 7600 kg. If it is also assumed that material and waste transport will be complementary to crew transport a minimum of four sorties by the larger submersibles presently available for dry transfer tasks (Taurus PC 1814 kg, DSRV PC 1950 kg) would be required every two weeks in a concentrated period, assuming a passenger complement of 25 crew per sortie.

The difficulties of handling such heavy submersibles from surface support vessels may be insurmountable, especially in adverse weather conditions. If alternative support systems are not available a trade-off study should be done between the acceptable number of sorties versus payload capacity. It may be better to use smaller submersibles more easily launched and retrieved more often, although this may compound docking and operational problems. There may also be good safety reasons for reducing the number of personnel exposed to potential accident situations during the transport phase.

10.4 DEEP SEA RESCUE VEHICLE, U.R.F. AND TAURUS

These three submersibles are at present the only vehicles that have the capability to give the payload capacity for our requirement for the support of a manned underwater structure. A brief review of these vehicles is now included to indicate developments in submersible vessels for support operations.

10.4.1 Deep Sea Rescue Vehicle (D.S.R.V.)

There are two D.S.R.V.'s, Avalon and Mystic which were built for the U.S. Navy's Submarine Development Group One by Lockheed Missiles and Space Company. The impetus for the construction of the submersibles came from the sinking of the nuclear submarine U.S.S. Thresher in 1963. The vehicles exist for the rescue of crews from disabled submarines at depths between the collapse depth of the pressure hull and free ascent escape depth limitations (see Chapter 11).

The submersibles are capable of mating with a 'downed' fleet submarine which may be listing up to an angle of 45°. The submersible has a crew of three and can carry 24 passengers (survivors). On docking with the distressed submarine the crew are transferred through the dry transfer hatch from the submarine into the D.S.R.V. and then transported to a mother submarine or surface support vessel. The procedure is repeated until all the crew are recovered.

The submersible has sophisticated sensor and navigation systems for search and docking roles. Three pressure spheres are encased in a hydrodynamically shaped G.R.P. hull. The forward sphere accommodates the pilot and co-pilot. The mid-sphere incorporates the transfer skirt. The survivors are carried in the mid- and aft-spheres (Fig. 10.4.1).[90]

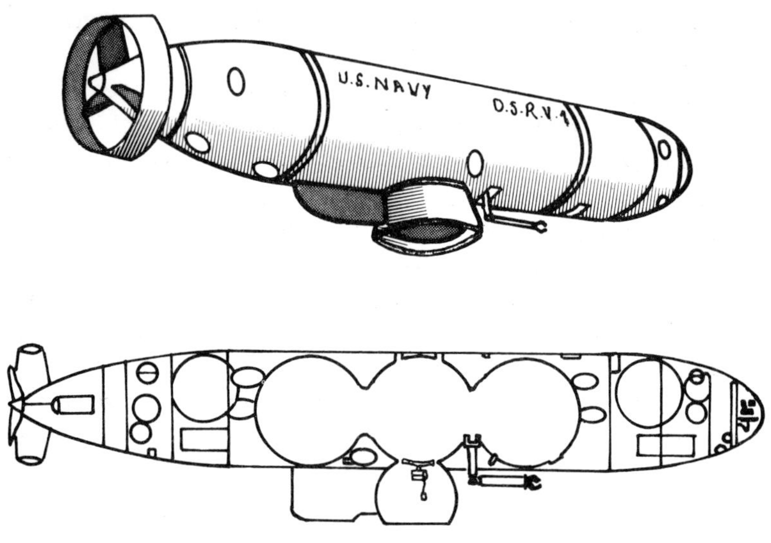

Figure 10.4.1 D.S.R.V.

The D.S.R.V. crew and supporting equipment can be transported worldwide aboard four C-141A or two C-5A jet transporters. A specially designed trailer for handling the D.S.R.V. overland is carried in one jet. The final leg of the journey to the disaster site can be achieved by riding 'piggy-back' on a submerged 'mother submarine', which serves as an underwater base launching and recovering the D.S.R.V. while submerged.

In 1979 during a test exercise the 'Avalon' and equipment were flown from the West Coast of the United States to Scotland, where it made transfers from one submarine (HMS Odin) to the mother submarine (HMS Repulse). The mother submarine was ready to leave the dock 33 hours after the operation started and it took four hours for the D.S.R.V. to mate with the distressed submarine after arriving at the disaster site.[96]

10.4.2 Submarine Rescue Vehicle (U.R.F.)

The submarine rescue vehicle (U.R.F.) has been designed and constructed for the Royal Swedish Navy by Kockums of Sweden. A civilian version of this vehicle could also be used for installation, maintenance and inspection operations in the offshore industry. The submersible has a maximum diving depth of 460 m and a payload capacity of 3000 kg. A diver lockout facility is incorporated in the vehicle for use to a maximum depth of 300 m.[91d]

The pressure hull is divided into four compartments. The forward operator's compartment contains all the controls and equipment for operating the vehicle, while the rescue compartment can accommodate 25 passengers. The auxiliary compartment contains electrical and gas outfits and distribution panels, further aft is the divers' compartment fitted with an entrance hatch and in the bottom an exit hatch (Fig. 10.4.2). The pressure hulls are encased in an external hull for hydrodynamic streamlining.

In a submarine rescue operation the vehicle is towed either on the surface or submerged by a surface vessel using a towing hawser,

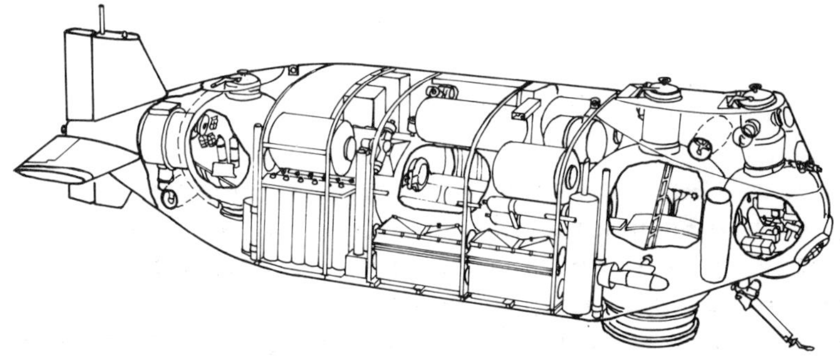

Figure 10.4.2. Submarine rescue vehicle U.R.F.

from the port to the disaster site. At the site the hawser is released and the U.R.F. homes in on the stricken submarine using sonar and a manipulator is used to secure a wire rope to the hatch cover of the submarine; divers can assist in this operation if water depths are less than 300 m. The submersible then hauls itself down to mate with the submarine and performs a dry transfer of personnel. The U.R.F. can be docked at angles up to 45°. After surfacing with survivors the submersible is towed back to harbour.

10.4.3 Taurus

The Taurus is a commercial submersible designed and built by International Hydrodynamics Limited (HYCO) of Canada. The large submersible has at present an 800 kg payload and can incorporate facilities for diver lockout, dry transfer into one atmosphere subsea chambers or can use a special mating skirt for the rescue of crews from military submarines. It is designed for the servicing of one atmosphere cellars and also offers a cost effective alternative to the sophistication of the D.S.R.V. for crew rescue.[91c]

The pressure hull consists of two basic sections, the command chamber and a dry transfer chamber. The command chamber has a 36 inch viewport forward and a removable conning tower with five viewports of 5 inch inside diameter. The command chamber is a 72 inch diameter stiffened cylinder with hemispherical ends. The cylindrical section is 80 inches long. The dry transfer chamber is an 84 inch diameter sphere and this is joined to the command chamber with a 27 inch diameter hatch opening. A double hatch is provided in the bottom of the dry transfer (D.T.) chamber and various mating skirts can be attached below the D.T. hatches (Fig. 10.4.3).

The Taurus hull, batteries, gas spheres and other major components are supported by an aluminium tube frame. Two aluminium skids are attached to the frame through shock absorbers and are provided for supporting the submersible. The skids are also retractable. When mating with a one atmosphere cellar the submersible lands on top of the mating adapter with skids extended. On centring over the mating flange the de-watering pump is started and the skids are retracted.

The operational experience of Taurus will provide valuable

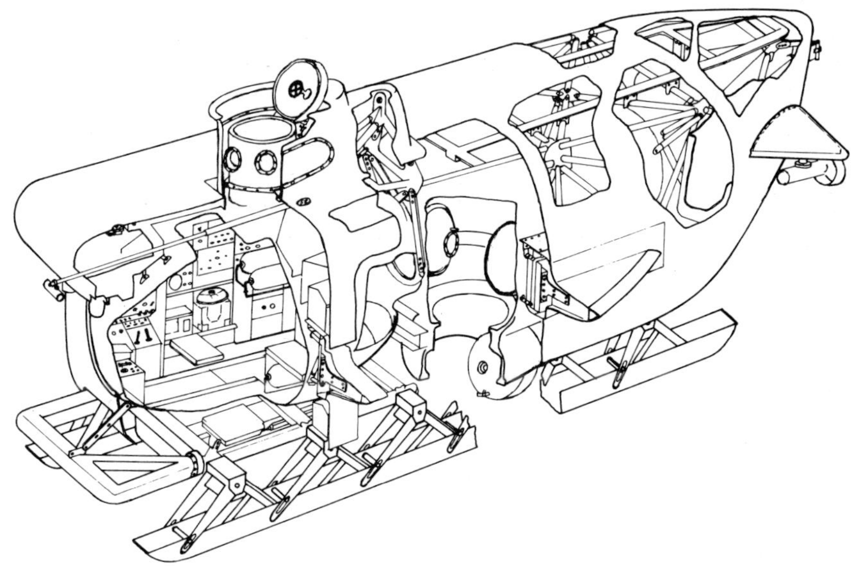

Figure 10.4.3 Taurus.

information on dry transfer operations to one atmosphere underwater structures and would be a more cost effective alternative to the D.S.R.V. for our requirements. The payload capacity is suitable, but its ability to carry large numbers of personnel needs to be further investigated.

The submersible required for our tasks would not need the sophistication of the D.S.R.V. Location of the structure will be relatively simple and it is most unlikely that the structure will have a significant list. Special equipment to assist mating of the submersible on the structure hatches can be incorporated in the structure design. The experience of D.S.R.V. operation and design should however be investigated in detail in conceptualizing a design of submersible for the support of the underwater structure. The use of large submersibles to obtain the payload requirement should be balanced against the difficulties of surface support operations. The use of alternative docking arrangements, subsurface bases and mother submarines should also be investigated.

10.5 THE MINI-SUBMARINE

The mini submarine is generally considered an autonomous submarine vehicle that is bigger than a large submersible but may be smaller than a conventional military submarine. To operate in an autonomous role the submarine must be able to travel to and from its place of work unescorted, recharge its batteries on the surface in all weather conditions, be totally self-supporting while at sea and capable of useful work. Crew sizes and payload capacity will be larger than those of a normal submersible.

The August Piccard (Ex-P.X.B.) has existed since 1963 and is a good example of a mini-submarine. Its diesel electric generators give it a range of 1000 nautical miles and it has an operating depth of 700 m. The payload capacity is 10 tons and it has a crew size of 6 with a capability to carry 4 observers or up to 10 passengers. The vehicle at present is being used for geotechnical and topographic surveys (Fig. 10.5.1).[90]

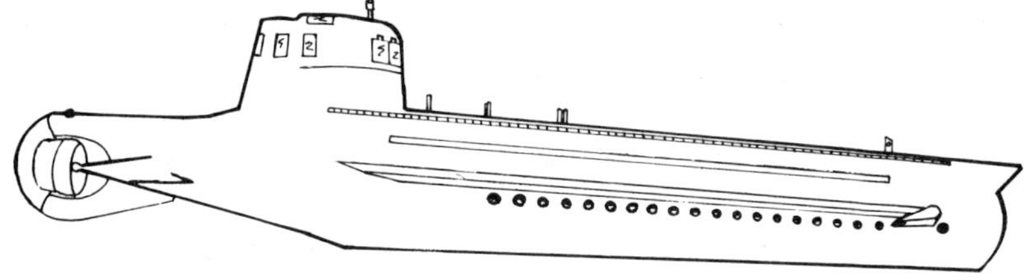

Figure 10.5.1 August Piccard.

Contemporary designs of autonomous mini-submarines appear to be limited to the work of Professor Gabler of Ingenieurkontor of Lübeck (IKL) in West Germany and Kockums of Sweden. IKL have designed the Tours 200/500, the smallest boat to meet the autonomous role, and the Tours 430/500 which is a longer boat built to similar parameters. Kockums were designing two mini-submarine types for full autonomous operation for the offshore industry. One was a 170 ton boat for inspection missions with an endurance of 10 days and the other was a 400 ton boat of mission endurance of 3 weeks and operating depth capability to 300 m. The Kockums designs, however, have now been shelved.

The Tours mini-submarines both have an operating depth of 500 m (Fig. 10.5.2). The Tours 200 would have an intermittent

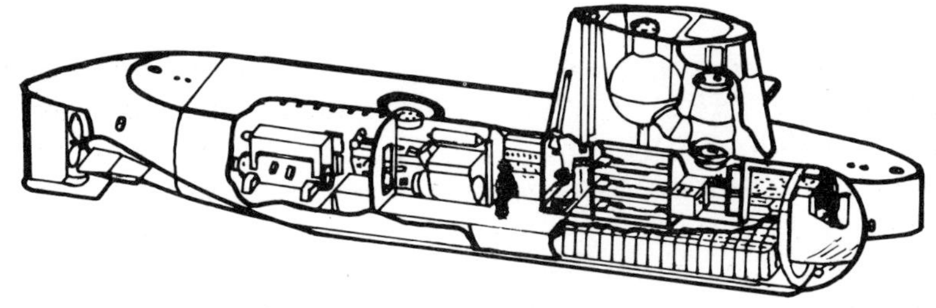

Figure 10.5.2 Tours 200.

endurance at 3 knots of 2000 nautical miles and 97 miles submerged at 3 knots and could operate from any port with containerized back-up facilities. A diver lock-out complex or mini-submersible could be carried abaft of the fin. The Tours 430 is similar in design but longer (42.5 m) and allows the fitting of a permanent diving complex and a core driller. The payload capacity is 10 tons and the battery capacity 6230 Ah.[97]

The autonomous mini-submarine used as a transport vessel to the underwater structure solves the problems of surface support systems and payload; however, it may not be possible to dock with the structure in the normal way because of the vessel size and its lack of manoeuvrability. The greater mass of the submarine could cause larger impact forces when docking, these may be alleviated by the use of shock bumpers. An alternative docking arrangement, such as the sea-lock mentioned earlier could be used or the mini-submarine could carry a smaller submersible for transport between itself and the structure subsea.

10.6 ATMOSPHERIC DIVING SUITS (A.D.S.)

The atmosphere diving suit (A.D.S.) (Fig. 10.6) is considered in this chapter as it may have application within our concept of an underwater structure for activities external to the structure at water depths greater than diving depths or in the emergency/escape situation. Atmospheric diving suits are articulated anthropomorphic or semi-anthropomorphic pressure vessels, which permit the support of underwater activities in water depths to 700 m, while retaining the single operator in a one atmosphere pressure environment.

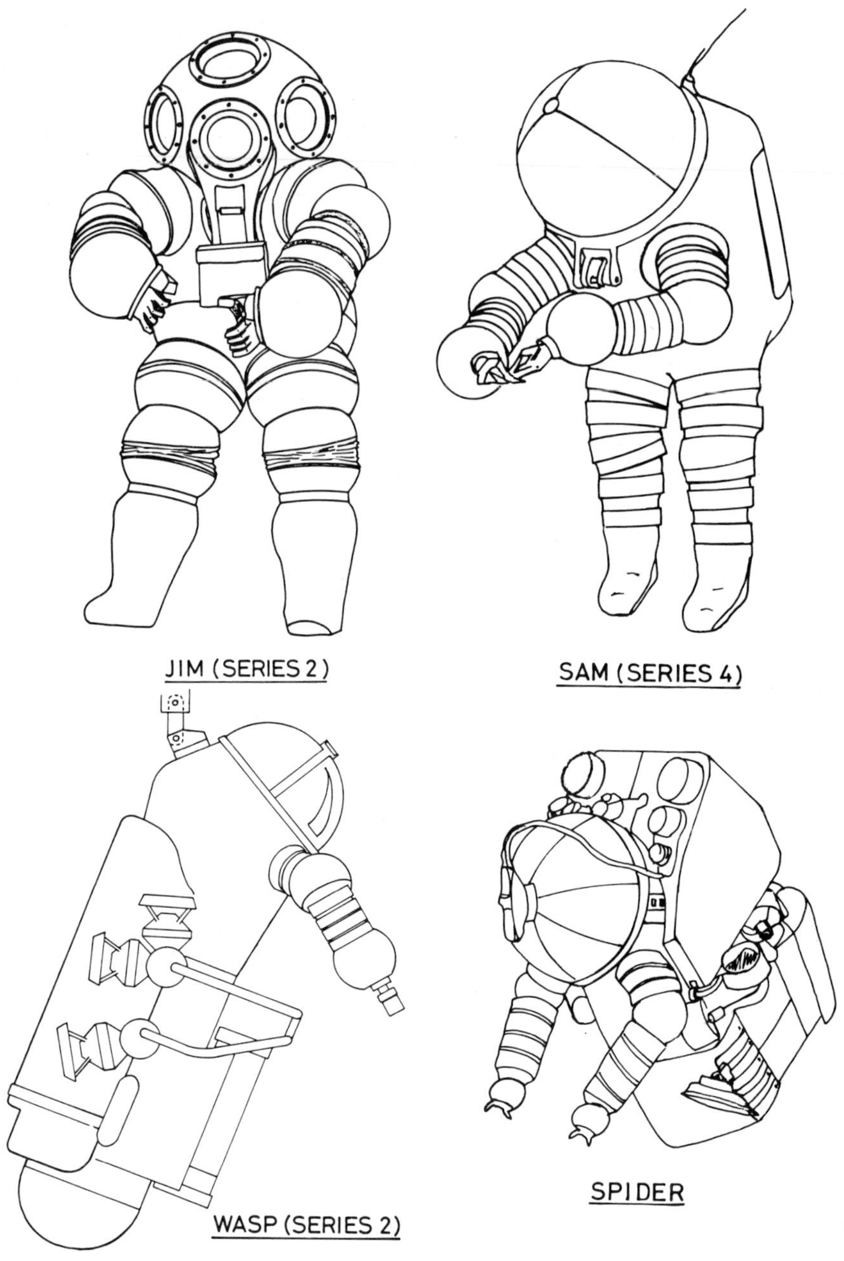

Figure 10.6 Atmospheric diving suits.

The fully anthropomorphic suits have a high degree of reliability because they are simple and directly operated by the human's limbs. The suits have a self-contained life support system and communications are incorporated in a lifting line. The JIM A.D.S. can be made of magnesium alloy, aluminium or glass reinforced plastic. JIM is fairly roomy with a dome entry, while SAM is smaller and hinged for entry at the waist. These G.R.P. models were produced using technology developed for submersibles and are also rated to 600 m.[98]

Semi-anthropomorphic suits, such as WASP retain the anthropomorphic character of JIM by virtue of its articulated arms and vertical mode of operation, however, horizontal and vertical thrusters replace the legs effectively giving it the capability of a minimum volume, one man submersible. Power is supplied to the motors via the lifting wire, although emergency battery power is

TABLE 10.6 Modern atmospheric diving suits *(Ref. 100).*

Name	*Builder*	*Characteristics*
JIM	DHB Construction	'Man-shaped' with four flexible limbs. Depth capability 400m. Suitable for bottom use only. Hand operated manipulators.
MANTIS	OSEL	Depth capability 1000m. Cylindrical, 8ft. in length, manipulator. Midwater capability with 8 thrusters.
SAM	DHB Oceaneering Int.	Similar to 'JIM' with more flexible joints and greater depth capability.
OMAS	Vickers-Slingsby	Depth capability 600m. Cylindrical body with articulated arms. Six thrusters.
WASP	OSEL	Depth capability 1000m. Cylindrical, 7ft. in length, articulated arms, power tools. Midwater capability with vertical and horizontal thruster.

provided for back-up. A recent addition to this group of A.D.S. is SPIDER, built by Slingsby Engineering for Wharton Williams, which incorporates articulated arms (Table 10.6).[99]

The JIM type systems are limited to sea bottom work and the seabed must be fairly solid and uncluttered, otherwise special platforms need to be constructed around equipment to allow access.

WASP type systems have evolved to satisfy mid-water requirements, but are somewhat restricted in the application of high forces and torques due to a lack of reaction to such forces. Many engineering tasks can, however, be undertaken satisfactorily.[100]

A lot of work experience has been gained by JIM and to a lesser extent by WASP systems. If greater attention is given to the design of subsea systems so that access and interfacing of A.D.S. to the equipment is improved, they can offer an extremely cost effective method of providing underwater support for subsea equipment in deep water. Suits having a depth capability of 1500 m are now under development.

The A.D.S. may be suitable for carrying out external work tasks, i.e. external inspection of the structure and servicing of peripheral equipment (satellite wellheads). A modified design may also serve as personal escape capsule in emergency conditions in areas of the structure that are remote from central escape systems.

11

ESCAPE AND RESCUE FROM AN UNDERWATER STRUCTURE

There are few types of accident that seem to excite the public interest as do submarine and submersible incidents. The reason is probably partly due to the helpless situation the public finds itself in, and the drawn out efforts required to rescue the crew from the situation. Public opinion, as well as humanitarian considerations, will require that the design of an underwater manned structure includes facilities for the rescue or escape of personnel from the complex in an emergency. The availability of rescue and escape systems will also encourage a sense of security amongst the crew while on station.

If it is assumed that the underwater complex will be strongly negatively buoyant and incapable of producing the positive buoyancy necessary to raise itself to the surface, the difficulties of rescue and escape will be similar to those of a 'downed' submarine. The various philosophies adopted by different Navies of the World for the rescue of submarine personnel can, therefore, be assimilated into this investigation of rescue and escape systems for the structure. Submarine rescue is, however, complicated by many factors due to its dynamic characteristics. The location, depth and attitude of the submarine may not be known initially, communications will be limited and surface support and rescue facilities will be remote from the disaster area. A permanent complex on the seabed will have a fixed location and be in a known water depth.

The structure will be sitting relatively level on the seabed and in communication with surface support or rescue facilities. The rescue of personnel from such a structure will be inherently simpler than submarine rescue. The incorporation of escape systems within the complex will not be restricted for space as they are in warship design.

In the escape situation, survivors of the initial incident will be located in an unflooded or uncontaminated compartment at a pressure that may vary from one atmosphere. The ultimate objective is survival, and the first task is to contain the consequences of the accident to gain time for the arrival of rescue facilities at the surface or on the structure. The internal design of the structure incorporating water-tight bulkheads, hatches, escape passages and safe havens will be critical to crew survival before escape and rescue is commenced.

This chapter assesses present methods for the escape and rescue of personnel from submarines. Buoyant ascent individual escape methods are limited to 183 m (600 ft), and because of the increasing collapse depths of operational submarines, rescue transport systems are being developed to cope with deeper submarine accidents.

11.1 ESCAPE

11.1.1 Individual Buoyant Ascent

Individual buoyant ascent from depth is the primary method of escape from Royal Navy submarines and provides an escape method to 183 m (600 ft). The escape system is independent of outside assistance except for pick-up on the surface, and avoids the problems of submarine location and maintaining a 24 hour service of rescue vehicles worldwide. The method caters for submarines 'downed' on the continental shelf, where the major portion of submarine accidents occur.

The system consists of special one-man escape towers within the submarine and a purpose designed immersion suit incorporating a buoyancy stole (Submarine Escape and Immersion Suit, S.E.I.S.) (Fig. 11.1.1a). The United States Navy have developed a modified version of the suit (E.A.S.E., Fig. 11.1.1b). The suit

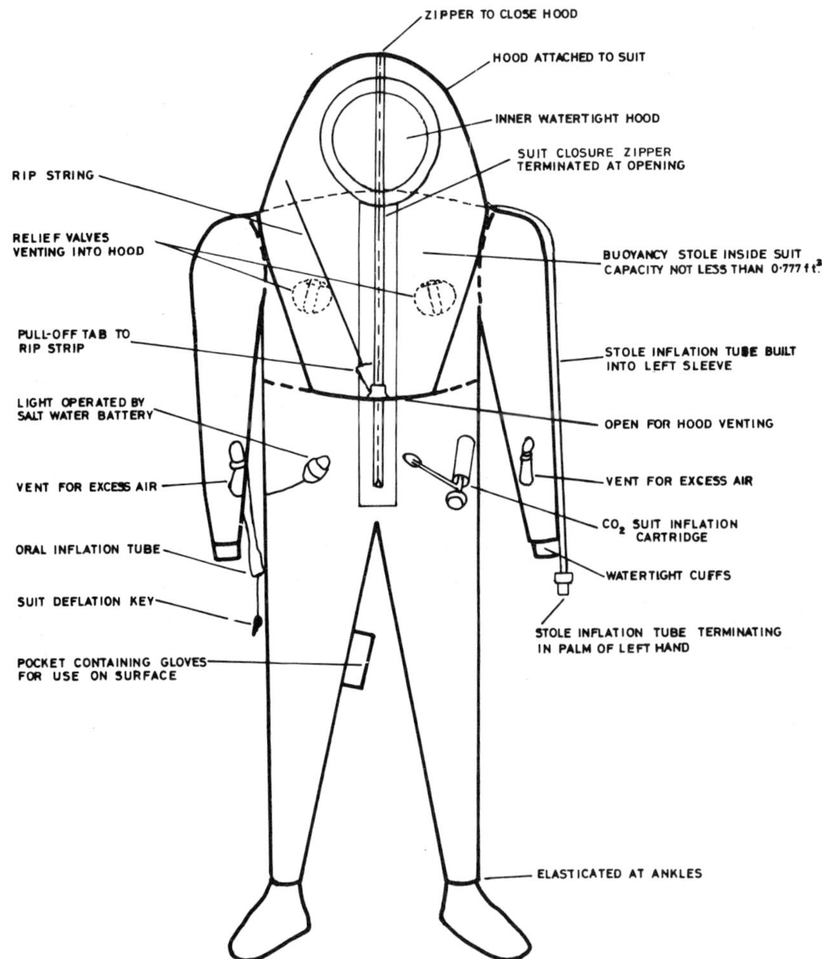

Figure 11.1.1(a) Individual buoyant ascent — British Mark VII submarine escape immersion suit (S.E.I.S.).

allows the survivor to breath normally during his ascent and protects him against the environment, while on the surface awaiting rescue.[103]

In the submarine, the survivor dons his suit and enters the one man escape tower through the lower hatch. He plugs his suit into a regulated and pure high pressure air supply, the air passes to the buoyancy stole and via relief valves to the hood, excess air being ventilated to the atmosphere by a valve. The lower hatch is then shut. The crew in the submarine open the flood valve to ambient

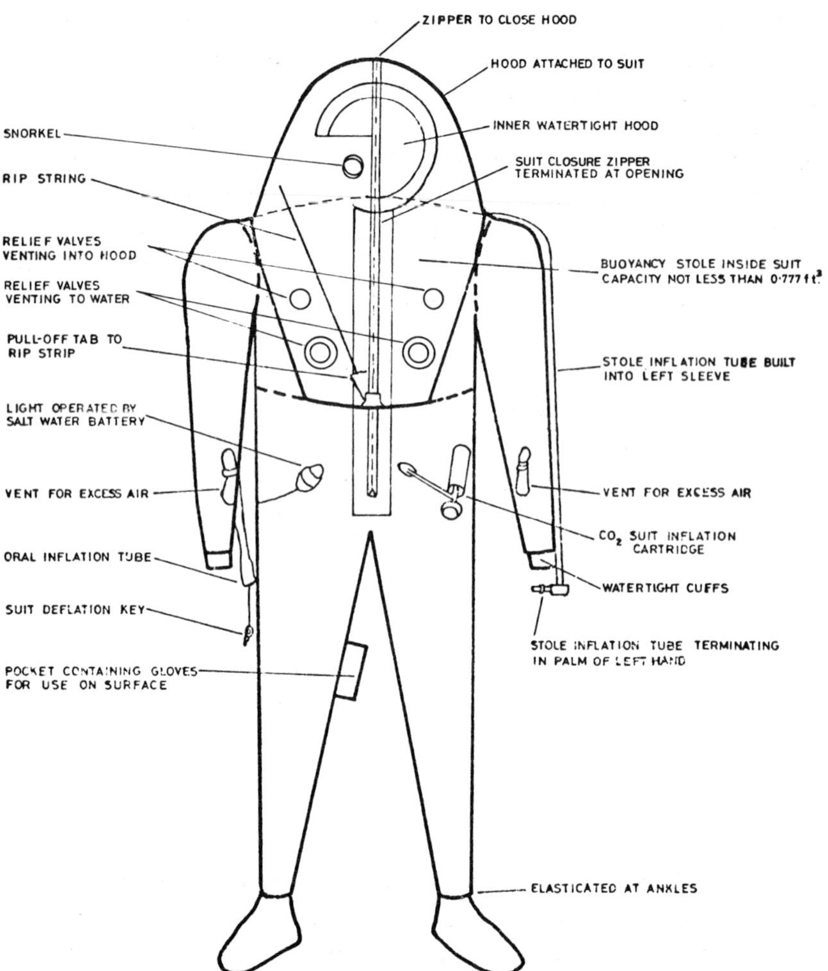

Figure 11.1.1(b) Individual buoyant ascent — escape and survival equipment (E.A.S.E.), Mark 1, Mod 0.

sea water pressure, water enters the tower displacing air out of the tower. The water level reaches the vent and the vent is shut and rapid pressurization occurs until the pressure is equalized to the ambient pressure. The pressurization phase takes less than 30 seconds. The escaper can then open the top hatch and releasing his air connection, he will be ejected by his buoyancy to the surface. Once sighted clear of the upper hatch, this hatch is closed and the tower drained down and the cycle repeated.[104]

The procedure is simple and requires a minimum effort from the escaper. The method depends on rapid compression to ambient pressure and a quick ascent to the surface to reduce the time under compression to a minimum to avoid the physiological effects of delayed decompression. Rapid compression is obtained by minimizing volume and using ambient sea pressure for pressurization, the ascent time being determined by buoyancy and escape depth.[105] Table 11.1.1 shows no-decompression times for various escape depths breathing air.

TABLE 11.1.1 No-decompression limits as a function of depth (air).

Depth (keel) (ft)	*Time (minutes)*
50	100
100	25
150	7
200	3.75
300	2
400	1.25
450	1
500	0.75
600	0.5

The possibility of extending the depth capability of this technique revolves around changing the composition of the breathing gas for the pressurization phase or some form of decompression procedure during the decompression stage. Experiments suggest that a worthwhile extension in escape depth may be achieved by the insertion of a decompression stop, although changes in breathing mixture would not seem beneficial.[103]

This system of escape would appear to only have potential application at the shallowest proposed depth of operation, 200 m. At deeper depths, it may be possible to adapt the concept of the one atmosphere diving suit into a personal escape system, which may be released on a buoyant ascent to the surface in a similar manner. Generally, as the location and depth of the structure are known, escape by buoyant ascent is most unlikely to be justified unless preferable alternatives are unavailable.

11.1.2 One Atmosphere Buoyant Ascent Capsules

A recent design for submarine escape is the concept of a rescue sphere developed in Germany by Ingenieurkontor, Lübeck (IKL). The present concept consists of a sphere of 2.1 m diameter, which provides seats for 22 people and room for 2 standing in the centre of the sphere (Fig. 11.1.2a).[107]

The sphere has two access hatches and can be entered from two compartments either side of a bulkhead (Fig. 11.1.2b). Each access leads through a hatch in the pressure hull of the structure and a hatch in the sphere. The pressure tight storage of the sphere on the hull is achieved by hatch coaming rings and sealings. The space between the hatches of each passage is drained into the structure for checking purposes and the seal is maintained by the

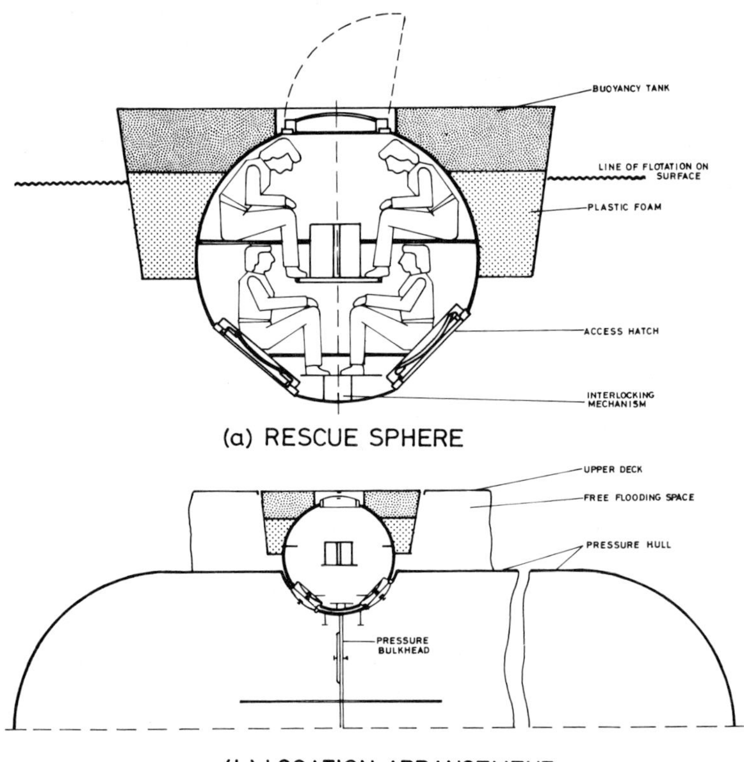

Figure 11.1.2 One-atmosphere buoyant ascent capsule.

hydrostatic pressure. At shallow depths where the external pressure is insufficient for the pressure effect, the sphere is locked in position by a mechanical locking device.

On transfer of the crew to the sphere for rescue, the pressure hull hatch is closed and then the sphere hatch. The intermediate space is flooded releasing the pressure and enabling the slipping of the sphere. In this procedure an unlocking and release mechanism is operated from within the sphere. The sphere is positively buoyant due to its own displacement and pressure-proof foam buoyancy arranged in an outer hull, which surrounds the upper section of the sphere. The buoyancy and shaping of the superstructure are arranged so that with a full complement of passengers, the sphere will always ascend to the surface and when there, will float with sufficient surface stability and freeboard to allow the upper hatch to be opened. Decompression can be performed before the hatch is opened. Life-support equipment is incorporated for the ascent and decompression phases and emergency rations and distress equipment are provided for surface survival.

Tests have been undertaken on models to assess the release mechanism, floating off the sphere at inclination and in currents, the behaviour during ascent, the behaviour in swell and under tow on the surface. The success of these tests indicates that the rescue capsule provides a means of escape from subsea vehicles and stations on the seabed.

The design could be modified and based on a cylindrical pressure hull to improve space availability. Such a design could be incorporated within the permanent undersea structure as a continuously available escape system and may also serve as a safe haven while awaiting the arrival of external rescue facilities.

11.1.3 One Atmosphere Diving Suits

The one atmosphere diving suit with its self-contained life-support system provides protection from ambient pressure and a contaminated environment. A modified design of present suits may provide individual escape capsules for buoyant ascent to the surface in the normal manner providing they are compatible with hatch dimensions.

An application may exist for these systems in relatively isolated areas of the structure, where crew numbers and intervention are restricted. Special escape suits could be made available for the limited crew in these areas, in case they become entrapped and isolated from other escape facilities. The suits could be used for life-support during contamination incidents and, in the extreme case of flooding, for protection against ambient pressure and for buoyant ascent to the surface. The concept will require a detailed design study to assess the limitations and difficulties of such an application (Fig. 10.6).

11.2 RESCUE

11.2.1 Submersibles

The use of submersibles for the transport of personnel and materials to and from the underwater complex has been discussed in Chapter 10. The submersible system selected would also provide the primary means of rescue of personnel from the structure. As we have seen Taurus, Mermaid, D.S.R.V. and the U.R.F. have been designed to offer the facility to rescue entrapped personnel from 'downed' submarines. The major limitations are the deployment and recovery of the submersible and docking arrangements to the complex. The submersible could be deployed from a surface support vessel,

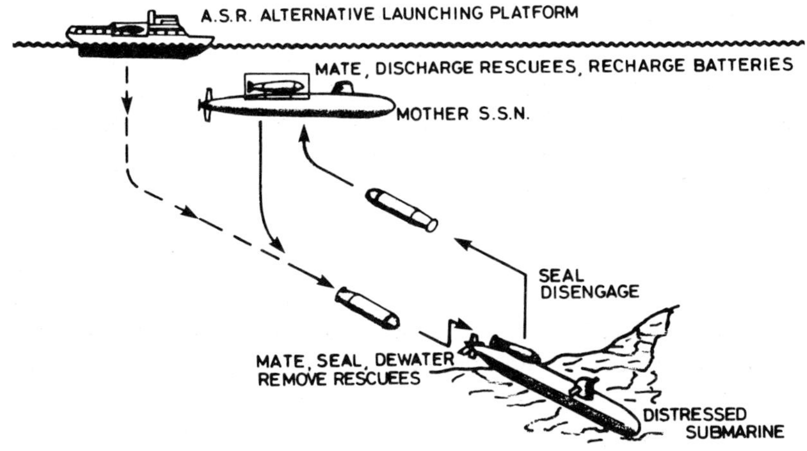

Figure 11.2.1 Rescue procedure-submersibles.

below the seawater interface of a surface piercing structure, or from an autonomous mother submarine (Fig. 11.2.1).

The use of submersibles for rescue will require the provision of access hatches on the structure that will provide a compatible mate to the submersible. The hatches should be provided at critical parts of the complex and standardized to worldwide submersible and bell rescue facilities. Guidance equipment should be provided to assist in the submersible's location of the hatches. The structure will probably incorporate safe havens under the hatches, where the crew can wait in safety and comfort until the submersible arrives and mates with the structure.

Submersibles have the potential to carry out rescue operations of personnel from an underwater structure. The method of deployment of the submersible in this role is a matter of debate and will be a complex function of other system parameters. The degree of surface support that is available or acceptable and the environmental conditions at the location of the structure will determine the choice.

The selection of a submersible support system must consider the requirement to operate as a rescue system as well as transportation system. Submersible operations will be the primary means of rescue of personnel from the structure.

11.2.2 Bells

The United States Navy originated the use of surface supported tethered bells for the rescue of personnel from submarines (McCann Bell). The present chamber (bell) operates to a depth of 260 m (850 ft) from an auxiliary rescue boat (A.S.R.). The A.S.R. lays a four point moor over the submarine and the positively buoyant submarine rescue chamber (S.R.C.) is attached to a cable on the submarine messenger buoy. The S.R.C. uses its air motor winch to haul itself down on the cable to the mating seal of the escape hatch of the submarine. The chamber consists of two compartments (Fig. 11.2.2), the lower open to the sea and providing the seal to the submarine hatch. The chamber carries off six survivors with the two operators. Difficulties are experienced with accurately mooring the A.S.R. above the submarine in deep water.

A new S.R.C. is being developed that will carry up to twelve passengers and a crew of two and operate at 457 m (1500 ft).

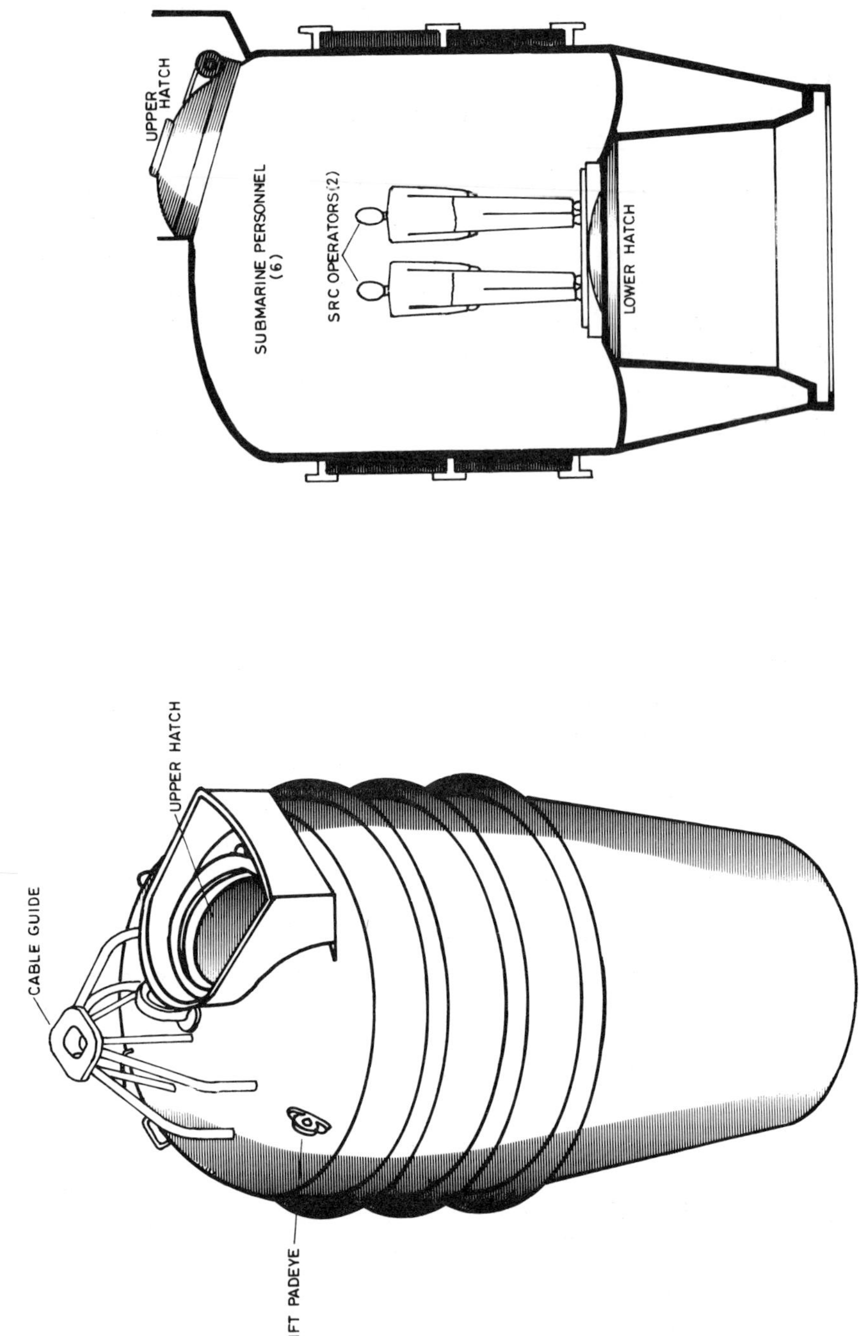

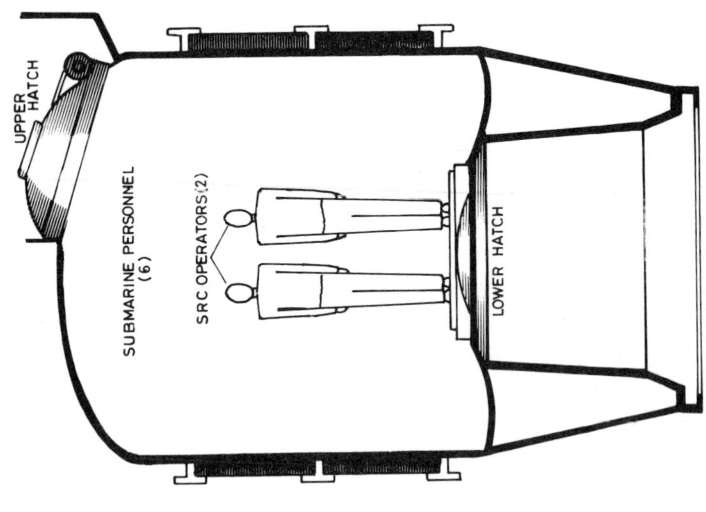

Figure 11.2.2 Submarine rescue chamber (S.R.C.)

A special S.R.C. fly-away kit has been developed for the 260 m S.R.C. that can be transported on three C-141's or two C-5A aircraft to the port nearest to the disaster site and then taken aboard a ship for rescue operations. A similar fly-away kit is being developed for the new 457 m S.R.C.[108]

At a fixed location and in the presence of surface support facilities, it may be possible to use this method of rescue for the crew. However, the dangers of anchoring over the subsea complex are obvious and should be avoided. It is unlikely that any advantage would be gained over submersible operations, which could be displaced from the area of the complex.

The deep submergence rescue vehicle (D.S.R.V.) is the primary means of submarine rescue, the development of the new S.R.C. provides the secondary submarine rescue capability of the United States Navy.

11.3 SUMMARY

The depth limitations of individual buoyant ascent and the logistic dangers of the use of submarine rescue bells make these unsuitable for the escape and rescue of personnel from our concept of a manned underwater structure. Submersible transport of personnel from the structure in combination with an acceptable support system is to be preferred as the primary rescue system. The incorporation of the one atmosphere submarine ascent capsule within the structure to act as a safe haven for submersible rescue operations and as a secondary escape system in the advent of the termination of submersible operations is recommended. The use of one atmosphere diving suits to extend escape depths by buoyant ascent could be considered for a tertiary level of escape, especially in isolated areas of the complex remote from central escape facilities.

The design of the internal structure of the complex will be critical to crew survival before rescue or escape is attempted. A comprehensive design study will be required of the layout of passages, water-tight bulkheads, hatches and safe havens in respect of escape and rescue facilities.

12

SUBSEA POWER SUPPLIES

The provision of suitable power supplies for the manned underwater structure will be a crucial element in the feasibility of the concept. Power will need to be supplied at two fundamental levels, a normal operating high power level (Megawatts = MW) and at an emergency level (Kilowatts = kW) in the event of the failure of the primary power system.

The selection of power systems to meet these requirements will be principally determined by the power level and endurance time. The depth of location of the structure and the remoteness of the structure from the shore or surface facilities will establish whether it is feasible to use an umbilical for the transmission of power to the complex from a remote atmosphere-breathing powerplant.

A power supply on the surface is obviously cheaper and safer to operate with the ready supply of oxidizer and ease of exhaust disposal, however questions of reliability are involved in the operation of transmission systems to the complex. Undersea cables from shore exhibit quite high breakage rates, especially on the Continental Shelf areas where they are exposed to strong currents and fishing activities. Surface terminated umbilicals experience physical disturbances due to weather conditions and strain due to their near vertical deployment.

In the conceptual design of a total subsea oil production complex it would appear fundamental to try and provide within the system an underwater power generation system to retain autonomous operation of the complex and reduce dependence on surface support facilities.

This chapter assesses the probable power requirements of normal and emergency operational procedures and relates them to present and possible future developments in power supply systems. The technology of air-breathing power systems for offshore operations is well-established, so the analysis concentrates on the more exotic power systems that have potential underwater application. A very detailed analysis is not possible in this book.

12.1 NORMAL OPERATING POWER REQUIREMENTS

Offshore oil production power requirements are strongly dependent on the configuration of the oilfields and the nature of the individual wells. Control power is always needed for the operation of valves, control systems and instrumentation, and will generally involve several tens of kilowatts of stable well-regulated supply. This power is normally obtained by battery storage and continuous recharging.

Stimulation for enhanced well flow and the pumping of oil to shore or a storage point requires much more power. Stimulation techniques include water injection and gas compression to pressurize the well or provide gas lift. Down-well electrically driven pumps may also be used to increase the yield of the well. The power requirement can range from hundreds of kilowatts to megawatts on multiple well fields.[109]

The pumping of oil over long distances to shore can also consume megawatts, rough estimates show about 100 kW are required per mile for each million barrels per day of production. Table 12.1 shows initial estimates of power consumption for an envisaged subsea complex in relation to a specific oil field configuration.[49]

Normal operating power of a production oil field can be of the order of tens of megawatts continuous.

TABLE 12.1 Power estimates for a subsea production complex *(Ref. 49).*

Field Production Operations	*Electrical Power Requirements (MW)*
1. Habitat Life Support only	0.60
2. Habitat + Basic Oil Production	3.17
3. Habitat + Basic Oil Production + Gas Export via pipeline and Fuel Gas	13.13
4. Habitat + Basic Oil Production + Gas Export via pipeline + Water Injection + Fuel Gas	19.40
5. Habitat + Oil Production (including NGL recovery) + Water Injection + Gas Injection	24.84

TYPICAL RESERVOIR DATA

1. Basic Oil Production = Oil transfer from separation to seabed storage + oil transfer from seabed storage to surface buoy + Tanker loading + Disposal of produced water
2. Production Rate = 100,000 brl/d stock tank oil (from 175,000 brl/d volume of wellhead fluids)
3. Tanker loading = 40,000 brl/hr
4. Produced Water = 20,000 brl/hr
5. Gas Oil Ratio = Volume of gas per barrel of stock, total gas production of 60 MMSCFD
6. NGL = Natural Gas Liquids
7. Stock Tank Oil = Crude Oil separated from other reservoir substances down to atmospheric pressure.

8. Gas Pipeline export pressure = 2200 psi
9. Gas Injection pressure = 5500 psi
10. Water Injection pressure = 2500 psi
11. Reservoir Pressure (Based on 3000m depth) = 4000 psi (approx)

12.2 EMERGENCY POWER SUPPLY

The emergency power supply, must be a non-atmospheric system independent of an umbilical connection and of the main subsea power supply unless modularization of the primary system can guarantee adequate partial power. The emergency power available must be sufficient to provide power for the crew to survive and be rescued. The power level and endurance of the supply will be dependent on the nature of the situation and the size of the crew. Basic requirements will include life-support, lighting, rescue and escape operations and possibly some control power.

Power requirements will be minimal for most of these functions, however, the need for crew heating under survival conditions will increase power levels significantly. A level of 10 - 20 kW will probably be required over a period of 5 - 7 days to maintain survival conditions before rescue or escape.

The emergency powerplant should be principally for emergency use and must be capable of passive stand-by for long periods (years) without loss of capacity and be capable of rapid activation, when the need arises.

An emergency power supply of tens of kilowatts is required with an endurance of 5 - 7 days.

12.3 ATMOSPHERE BREATHING POWER SYSTEMS

Atmosphere breathing powerplants can be considered if an electrical umbilical attachment between the sea-bottom structure and a surface facility is permissible. The use of diesel and gas turbine generators in the offshore oil industry for the generation of electrical power is well established. Submarine cable technology is available, although improvements in underwater 'wet' connectors are required. Submarine power cables can be laid for many miles and to depths down to 6000 m. The cost of these cables is not strongly dependent on thc power level, but they are affected by the costs of operation of lay barges and continuous maintenance.

The reliability of transmission cables and umbilicals may prove to be a limitation to their use with an undersea oil production complex, whose operation is sensitive to unscheduled downtimes. Umbilicals deployed in the vertical mode to the subsea complex will share many of the problems associated with deepwater riser technology, i.e. drag, vibration etc.

A detailed discussion of atmospheric power sources is not considered relevant here, but it is appreciated that cost and feasibility comparisons of these systems with alternative subsea sources will need to be undertaken.

12.4 NON-ATMOSPHERE BREATHING POWER SYSTEMS

A non-atmosphere breathing powerplant does not make direct or active use of the atmosphere and requires logistic disposal of the waste products. Subsea powerplant characteristics will be similar to those of a space station or lunar base. Batteries, fuel cells, radioisotopic generators and nuclear reactors can be considered in this category of powerplants. The independence of these systems from an atmospheric supply of oxygen makes them ideal candidates for the subsea provision of power. A review of these types of power systems is undertaken to establish the potential of these systems within the concept requirements (Fig. 12.4).

12.4.1 Batteries

Generally primary batteries will be better suited than secondary batteries for underwater operation, because of the difficulties of recharging and the ease of replacing the primary battery. The Lead/Acid battery, however, can be considered as a primary

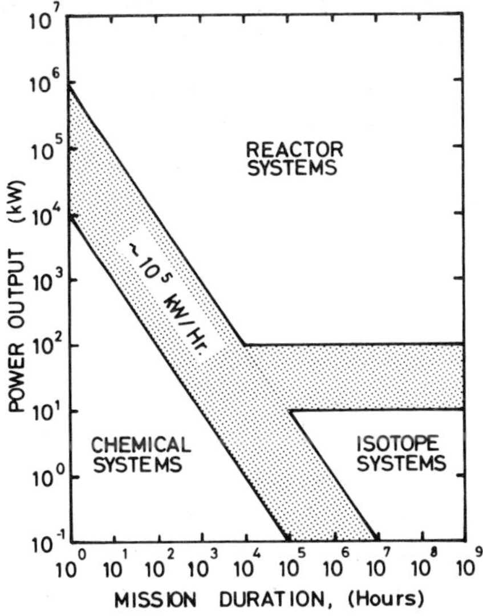

Figure 12.4 Undersea power systems.

battery because of its relatively low cost and extensive use for marine applications.

Batteries can only be considered for use in the emergency power system. All the batteries except the Lead/Acid and Silver/ Zinc (Ag/Zn) have not the power capacity to discharge over several days under survival conditions. Nickel/Cadmium batteries may need to be avoided as there is at present no reliable way of measuring their state of charge. Lithium thionyl chloride ($LiSoCl_2$) batteries developed in recent years by Altus in California offer potential for marine applications and could also be considered.

Silver/Zinc batteries are approximately ten times the cost of Lead/Acid batteries and have only one tenth the cycle life; however, they can be float charged on the main power supply for many years without gassing or electrolyte depletion. Lead/Acid batteries have poor 'wet stand' characteristics and must be kept on trickle charge to ensure full capacity, resulting in gassing and water loss. Lead/Acid batteries can be 'dry-stored' for long periods with automatic fluid insertion on main power failure, although there is a short delay (minutes) before the power comes on. A new battery development that shows promise is the Nickel/Hydrogen battery which uses a gaseous H_2 fuel in its own pressure vessel, the hydrogen pressure gives an ideal measure of the state of charge and it operates at ambient temperatures.[111]

For emergency use Ag/Zn or 'dry-stored' Lead/Acid batteries may be suitable, and in the future possibly the Ni/H_2 and $LiSoCl_2$ batteries. Other batteries do not have sufficient power capacities or have operational problems. Metal hydride systems for stand-by application are not appropriate because of self-heating requirements.

Storage volume and logistic supply problems will probably restrict the provision of batteries for the power level and endurance required.

12.4.2 Fuel Cells

The fuel cell is an electrochemical cell into which reactants are supplied externally in liquid, gas or solid form. The electrical generation system is not truly non-atmospheric as an oxidizer is supplied as one of the reactants; however, chemical energy is turned directly into electrical energy at an efficiency of 50 – 70%. The operating system can be very simple and involves no moving parts providing good reliability with very little operating attention.

Considerable research has been undertaken into the development of fuel cells in the last three decades. The major manufacturers today are United Technologies Corporation (ex Pratt and Witney), Alsthom in France, Westinghouse Electric and several automobile companies. Initially, fuel cells were very expensive power packs developed for military and space missions. General Electric developed a 1 kW power module for the Gemini space programme (Fig. 12.4.2) and United Technology a 2 kW module for the Apollo space vehicle, operating on hydrogen and oxygen reactants (HYDROX cell). Alsthom in France are developing fuel cells specifically for underwater application based on a 10 kW module and using hydrazine and hydrogen peroxide reactants. A zinc/oxygen fuel cell has replaceable anodes, which are expensive but exhibit long life expectancy. Recent developments in fuel cells allow the use of air as the oxidant and the reformation of hydrocarbon fuels, for utility applications producing megawatts. A regenerative type fuel cell offers similar characteristics to a secondary battery.

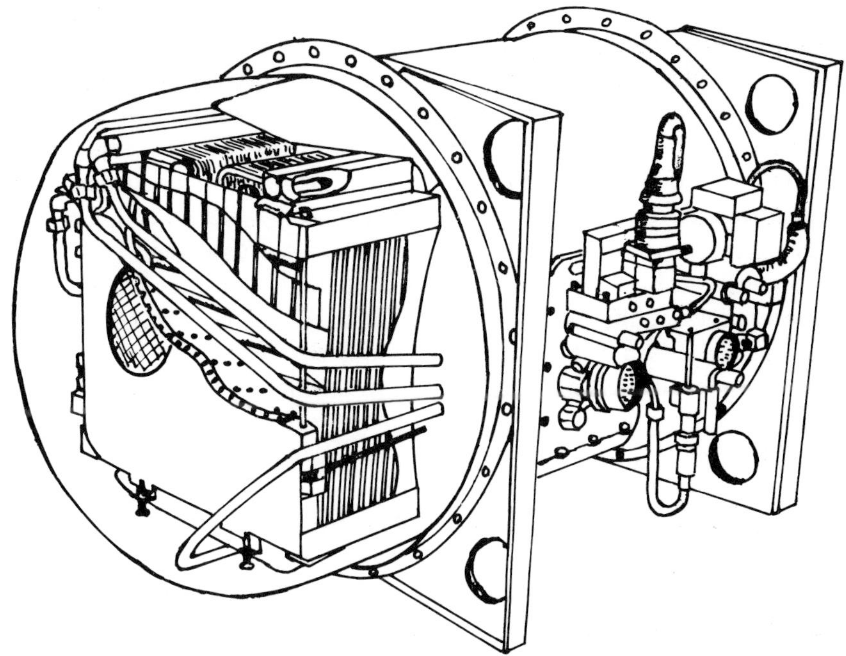

Figure 12.4.2 Gemini 1kW fuel cell powerplant.

United Technologies are producing for the United States Navy and N.A.S.A., fuel cells using pure reactants of oxygen and hydrogen and an alkaline electrolyte producing power in the range 10 - 30 kW with an efficiency of 70%, the reaction products being water and waste heat.[117] One of these systems was recently tested in the United States Navy's 'Deep Quest' submersible and produces 700 kWhr of 120V D.C. electricity for a weight of 1375 lb and volume of 6 cuft, neglecting fuel storage.[118]

Also under development by United Technologies is the use of fuel cells to develop power in the range 40 kW - 4.8 MW for utility applications in the electricity industry. The smaller units are for on-site installations and the large units for dispersed generation installations. A phosphoric acid electrolyte allows the use of the more stable hydrocarbon fuels and an air oxidant, the products of the reaction include CO_2 and the efficiency is of the order 45 - 50%.[119]

Fuel cells after many years of development seem to be coming of age and could make a major contribution to underwater power supplies in the 1980's. Fuel cells are unsuitable for the large power requirement because of the need to supply oxidant; they would be most suitable for emergency power because of their reliability and endurance. The membrane type hydrox cells would appear most appropriate as they operate at room temperature and experience little start-up delay, unlike similar systems which need to bootstrap themselves up to working temperature (1 - 2 hours). Oxygen and hydrogen would have to be stored as a pressurized gas, cryogenic storage because of boil-off would be unsuitable for a standby application. Zinc/Oxygen fuel cells may also be suitable as they do not exhibit a delay on start-up and the restrictive cost of the anode replacement kits would not be a problem for this application. Hydrazine/Hydrogen peroxide fuelled unit and the Lockheed Lithium/Sea Water cell which uses ambient sea water as the oxidant, may also be suitable.

An added advantage of fuel cells in the emergency situation is that the hydrox system produces water and waste heat as products of the reaction, which could be used to assist life-support functions in the survival situation and reduce electrical demand.

12.4.3 Radioisotopic Generators

Radioisotopes are materials which spontaneously emit electromagnetic radiation and/or nuclear particles. The energy generated by the radiation and nuclear particles can be utilized to produce thermal energy, and in some circumstances, to produce electrical energy directly. The energy production decays exponentially with time at a rate which cannot be altered physically.

Radioisotopes occur naturally and are also produced artificially in nuclear reactors, only the latter are suitable for electrical generation because of the high power density required to generate sufficient thermal energy. The thermal energy is converted to usable power by thermo-mechanical converters or by thermoelectric generators.

The selection of a radioisotope fuel for a specific application is determined by many factors. The power density of the source and the half-life of the material needs to be matched to the power requirement and endurance for the application. Cost, availability and ease of encapsulation of the fuel and production of the source have to be considered in the design. A significant part of the design and weight may be involved in protecting the environment from high radiation levels, with suitable shielding. These combinations of factors generally restrict their application to specialized situations where extreme reliability and long-term use are called for.

Fuel costs dominate the cost of a radioisotopic generator, so a highly efficient power conversion system is required to reduce the fuel inventory. Thermo-electric generation has high reliability and minimum maintenance but only operates at about 6% efficiency, thermo-mechanical systems offer 20% efficiency, but at present with reduced reliability.

Present developments appear to be taking two complementary directions. One is the use of high cost fuel (Plutonium) to produce low weight systems (SNAP 19) and the other is the use of relatively

SPECIAL FEATURES

* no fuel resupply
* long term unattended use
* non-pollutant
* kilowatts of thermal power for heat, water/waste function
* bacteria kill capability

CHARACTERISTICS

* 2-10 kilowatts
* 1 watt per pound
* efficiencies to 25%
* 2-5 year life
* ^{60}CO heat source for thermal use and various static/dynamic convertors
* Organic Rankine — 10kW Shielding — 7000 lb.

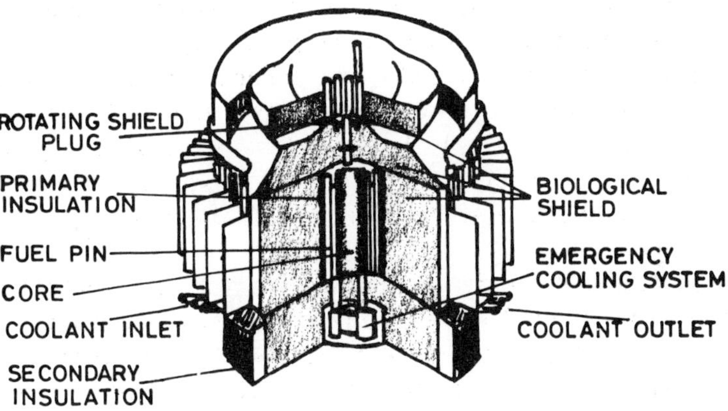

Figure 12.4.3(a) Large cobalt 60 heat source.

low cost fuel (Cobalt) and high weights, because of shielding requirements (Fig. 12.4.3a). Developments with ^{60}Co heat sources in combination with organic Rankine engines offer power in the range 2 – 10 kWe at an efficiency of 25%. An interesting development in space life support systems is the Lockheed R.I.T.E. programme (Radioisotopic Thermal Energy) in which the thermal energy source (Promethium/Plutonium) is used directly for life support heating and waste treatment to minimize the onboard requirement for electrical power.[120]

Previous conceptual studies of manned underwater structures have envisaged the use of radioisotopic power sources. General Dynamics[53] proposed a ^{60}Co source driving a steam turbine and A.C. electrical generator. The Westinghouse concept[52] proposed four separate power units using a ^{60}Co heat source and superheated Rankine cycle for power conversion each producing 7.5 kWe (Fig. 12.4.3b).

Radioisotopic generators offer a power range from several watts to several kilowatts at present, the efficiency of the system is determined by the method of energy conversion. Thermoelectric conversion is more reliable than thermo-mechanical

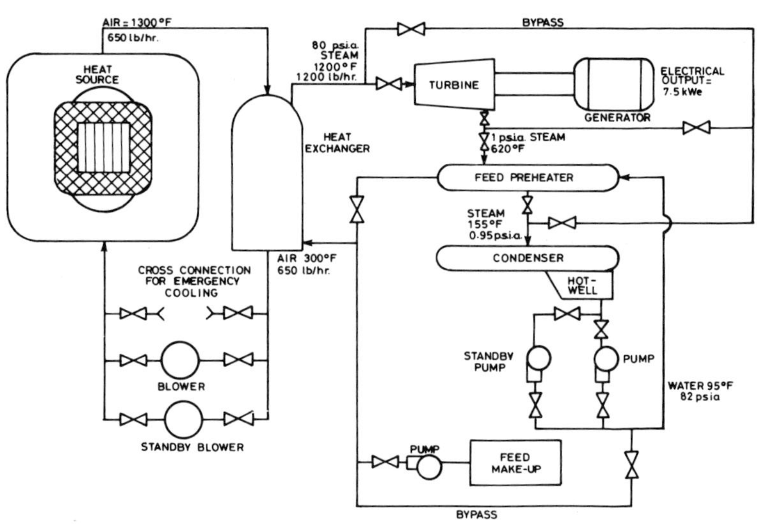

Figure 12.4.3(b) Schematic 7.5kWe power plant module — Westinghouse.

converters but less efficient. Developments in relatively low cost ^{60}Co systems utilizing thermo-mechanical conversion systems will produce electrical energy in the order of 10 kWe. These have previously been considered suitable for use in manned underwater structures.

Radioisotopic power sources may have application in our concept for emergency power or for situations where minimal power is required for field operations. The power would be continuously available and could provide a level of redundancy to the primary power source, as well as acting as an emergency power supply. High radiation levels associated with the power source could be utilized for the irradiation sterilization of waste products of the complex, and the cooling water incorporated into the emergency life support system for heating purposes to reduce the electrical demand.

12.4.4 Nuclear Reactor Power Plant

Nuclear fission has been utilized for power generation for several decades. The fission, or splitting of heavy atoms, produces two separate nuclei with the emission of intense gamma radiation and two or three neutrons having kinetic energy. A suitable arrangement will cause a chain reaction to be sustained. In such a nuclear reaction a change in mass occurs between the original nucleus and the masses of the constituent parts. This so-called Mass Defect is related to the vast amount of energy that is released on fission.

A reactor powerplant consists of a nuclear reactor for the fission process, this acts as the thermal energy source and is coupled to a suitable energy conversion system. Unlike radioisotopic power systems there are no theoretical power limitations and power costs only increase slowly with power level, also reactors are cold and inactive before start-up and can be shut-down to relatively low levels of thermal and radiation activity at any desired time. Protection against radiation levels by shielding is, however, still required.

In today's reactors the energy from the chain reaction is used to heat a coolant. The coolant is generally a liquid such as water or may be a gas or liquid metal. The hot coolant is often used to generate steam for use in a Rankine cycle engine or it may be used to provide heat energy for a Brayton cycle. Other methods are also used and a variety of reactor concepts can be coupled to various power conversion systems to provide a full range of performance and life capabilities. Various reactor and power conversion systems are shown below:

REACTOR TYPE	*POWER CONVERSION SYSTEM*
Pressurised Water Reactor (P.W.R.)	Water/Organic Rankine
Boiling Water Reactor (B.W.R.)	Brayton
Gas Cooled Reactor	Thermo-electric
Liquid Metal Fast Reactor	Thermionic
Thermal System	

12.4.4.1 Reactors

The pressurized water reactor (P.W.R.) is probably the most widely used and well developed reactor type; water serves as the coolant, moderator and reflector. The P.W.R. has been used extensively for land based electricity generation and in ship and submarine applications of nuclear power. The boiling water reactor (B.W.R.) has no secondary loop and requires careful control of the single cooling loop. Gas cooled reactors generally use CO_2 as the coolant and have been used for land based electricity generation for several decades, but suffer the disadvantage of low heat transfer. Liquid metal reactors use coolants such as liquid sodium and potassium metal giving compact reactors with excellent heat transfer properties at low pressures, however, they experience high induced radioactivity and a need to keep the

TABLE 12.4.4.1 Core sizes for different reactor types.

Reactor Type	*Core Volume (cuft)*
P.W.R.	200
B.W.R.	650
Liquid Metal	20
Gas Cooled	2000

metal molten at all times with external heaters during shutdown. Table 12.4.4.1 indicates core sizes for various reactor types.

12.4.4.2 Power Conversion

Thermo-electric power conversion is used in conjunction with a liquid metal cooled reactor in the SNAP 10A and SNAP 8 marine power system programme. Heat from the reactor core is rejected through thermo-electric elements giving an efficiency of 6 - 10% and powers of 5 - 50 kW with high reliability and long endurance. A new development in the SNAP 10 programme is the use of a solid-state reactor in which heat is conducted by metal plates to the thermo-electric elements replacing the liquid coolant; 1 kWe is generated at 4 - 6% efficiency.

Improvements in dynamic conversion systems are centred on the incorporation of hermetically-sealed combined rotating units. Light weight high speed turbo-electric equipment is mounted on a single shaft with process pumps, and automatically controlled. The aim is to produce a system capable of unattended operation for several years between overhauls. The B.W.R. is inherently simpler in operation but requires special provisions to ensure reactor stability and the gas cooled reactor has a relatively low reliability because of the high temperatures of operation and problems with bearings.

The organic/water Rankine engine conversion systems appear to offer the best potential for high power levels, hundreds of kilowatts to megawatts, as the efficiencies are better at these power levels. Thermo-electric systems are more reliable and competitive in the lower tens of kilowatts range.

Developments in the SNAP nuclear reactor programme, in the United States should be evaluated in further detail to assess progress in the application of nuclear reactors to terrestrial and oceanic functions in the tens of kilowatts to multi-megawatt range.

12.5 SUBSEA NUCLEAR POWERPLANT STUDIES

12.5.1 Rockwell Subsea Oilfield Reactor

A conceptual study was undertaken by Atomics International, a

Division of Rockwell International in association with Lockheed Petroleum Services and Exxon Production and Research, into the development of a small 3 MWe reactor for future offshore oil production systems (1977).[123]

It is claimed that the system could be designed and built with minimum additional development based on space nuclear technology components and systems completed in the last twenty years. The system is designed to produce 3 MWe of power for periods of four years, in an automatic, unattended mode. After four years the system is returned to a shore base where the reactor core is replaced and the system inspected and overhauled. The total system design life is 20 years.

The nuclear reactor power system is largely contained within two separate pressure vessels, connected by an enclosed pipe chase. One vessel contains the reactor and liquid metal primary coolant and the other contains the organic Rankine power conversion system. The vessels are mounted on a barge which provides a means of transport for the system and a mounting structure for the seafloor installation. A waste heat exchanger is also mounted on the deck of the barge. The system is designed for unattended operation and can be operated automatically in the load-following mode or controlled remotely with supervisory control systems (Fig. 12.5.1).

The heat core is zirconium hydride and this is cooled by liquid metal (NaK) coolant pumped by A.C. helical induction pumps. The organic Rankine system uses toluene as the working fluid, and the waste heat exchanger fluid is an ethylene glycol/water mixture.

The reactor generates 10 MW of thermal power and with a regenerative cycle exhibits an overall system efficiency of about 30%, giving 3 MWe power.

It has been estimated that 80,000 barrels a day of crude could be pumped over distances of 100 to 200 miles in 14 to 16 inch pipes, and raw production fluids having gas to oil ratios of about a thousand, over 25 to 50 miles at reasonable pipe sizes. Downhole lifting to a field of about twenty wells could be provided. The 3 MWe system gives a reasonable size and weight for installation and existing offshore lifting facilities could handle the envisaged weights.

The provision of the Lockheed barge, seafloor base and subsea

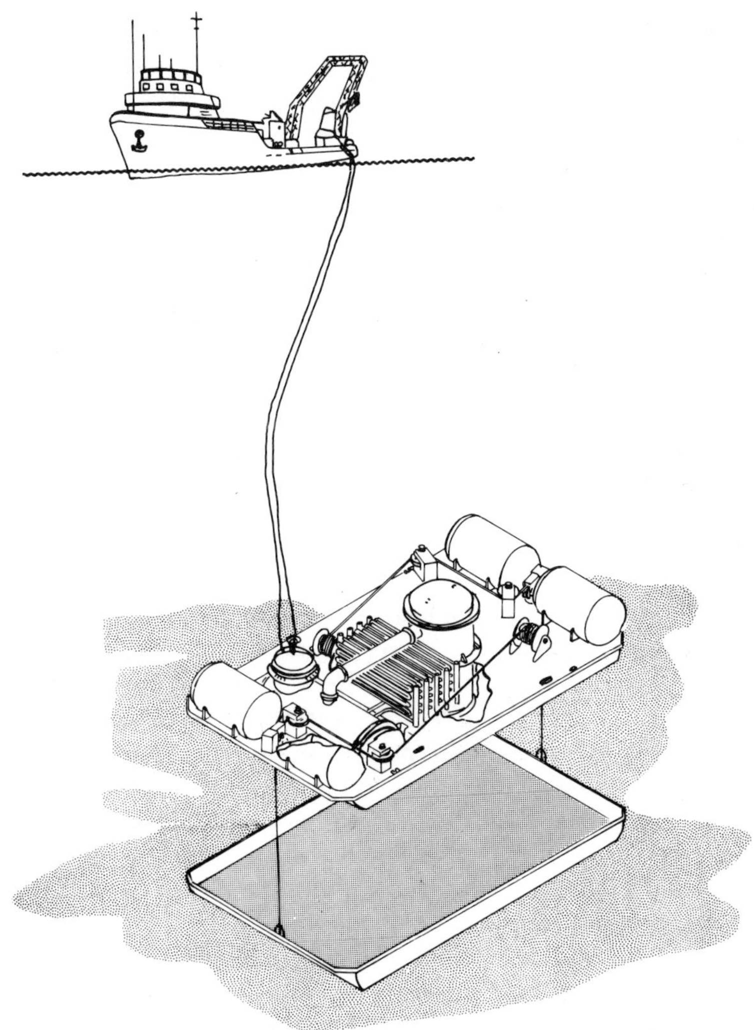

Figure 12.5.1 Rockwell subsea oilfield reactor.

installation make a substantial contribution to the costs, estimated in 1975, based on cost indices at that time, with no provision for the effects of subsequent cost inflation factors, as an installed cost of $10,480,000 with an annualized cash flow of $2,356,000 (13 cents/barrel). The Exxon application would install the unit on the main template of the Exxon Submerged Production System (S.P.S.) giving reduced costs of $6,875,000 installed and annualized cash flow of $1,873,000. The S.P.S. system includes gas-oil separator and the oil is pumped as degassified crude, with production rates of 80,000 b/d and pumping distances of 100 to 200 miles, it gives a unit cost per barrel of 6.4 cents.

12.5.2 Subsea Pressurized Water Reactor D.S.P.S. (Rolls Royce)

The conceptual design study of the subsea oil production complex by the D.S.P.S. Consortium considered the use of a subsea nuclear powerplant. Rolls Royce proposed the use of a pressurized water reactor with a dual turbogenerator combination in tandem with the single nuclear steam supply. The technology is based on previous operating experience in marine and land-based applications. The concept envisages a submarine nuclear power station (S.N.P.S.) docked for several years to the manned underwater structure.[124]

The deep sea production complex would be provided with three or more docking points for the nuclear power modules. The power requirement would determine the number of modules required. Multiple modules are required to give suitable redundancy within the power generation system and to allow for refurbishment and refuelling of individual power modules as required, while maintaining a continuous power supply. The S.N.P.S. modules could be transported and deployed by a specially designed surface support ship or, if environmental constraints are severe, the module may incorporate propulsion and steerage systems to allow autonomous operation.

The power modules would be continuously manned, access being provided from the subsea complex. Life support and rescue facilities will be available within the power module. Emergency power for reactor control and life support of personnel will be provided by batteries.

Power requirements for the complex are assumed to be 25 MWe and it is considered that to provide this power with a high level of confidence, a dual power source, i.e. two power modules, will need to be connected to the seabed complex. Each power source will be connected independently and provide 27.8 MWe, 25 MWe for operating systems and 2.8 MWe for reactor systems and auxiliary power. The power modules would normally operate at half-load, so that with planned *in-situ* maintenance of one module, non-interrupted power can be supplied by the other module at

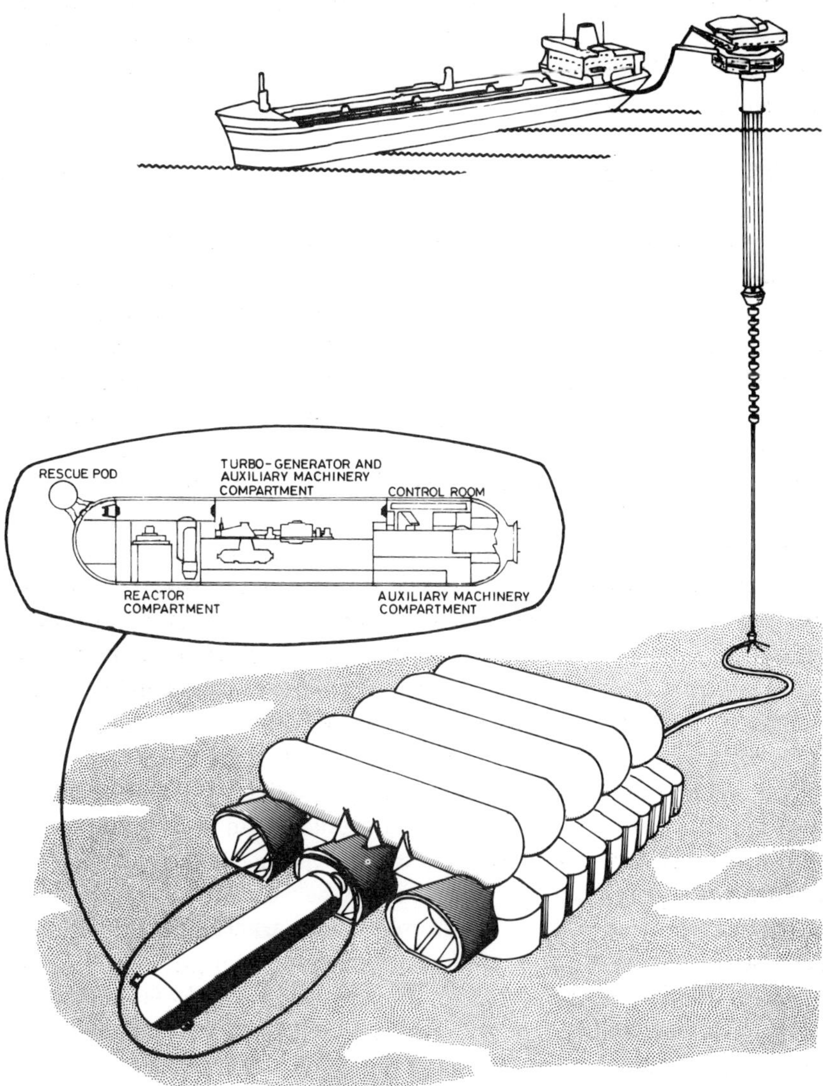

Figure 12.5.2 Seabed power generation (Rolls-Royce).

full-load. Periodically modules will need to be returned to base for refurbishing or refuelling, so a third reactor module is required to cover these events.

The normal core-life of the reactor is three years. However, normal operation at half-load will give an extended core life of six years, and with three modules, each module would only need to be refuelled every eight years. Total life expectancy of the system is twenty-five years, so two new cores for each power module would be required during its life. A biennial maintenance and refurbishing period should be suitable for servicing auxiliary systems. The dual-operating configuration will also allow shutdown of one reactor system for *in situ* inspections to satisfy regulatory authorities.

Figure 12.5.2 shows an artist's impression of the interface of the power modules to the complex and the power module layout.

12.5.3 Minimum Critical Mass Reactor (University of California, Los Alamos Labs, N.M.)

A pressurized water reactor suitable for unattended operation on the seafloor has been conceptually designed, which is claimed to have inherent safety, simplicity and low cost. The design is based on the principle of a minimum critical mass, such that any change in size, shape, materials or fuel density results in a less reactive condition and if the change is large enough it will cause automatic shut-down of the reactor. The reactor operates in a restricted zone where it is most reactive, any change produces less reactivity (Fig. 12.5.3).

The reactor core is a honeycomb structure of zirconium alloy and uranium at 93.4% $^{235}UO_2$. The core volume determines the maximum power level, and to achieve high powers a fuel of lower enrichment can be used, which increases critical mass and core volume even more, while retaining the characteristics and safety features of the multi-minima design.[125]

Thermo-electric elements would be used for power generation up to 15 kWe and dynamic Rankine systems for higher power levels. The unique safety features of the design appear to have advantages for unattended subsea operations.

12.5.4 Summary

Nuclear reactor powerplants can produce electrical power efficiently in a range from tens of kilowatts to multi-megawatt, however, in the lower range nuclear reactors will not be competitive with alternative systems i.e. radioisotopic generators, fuel cells.

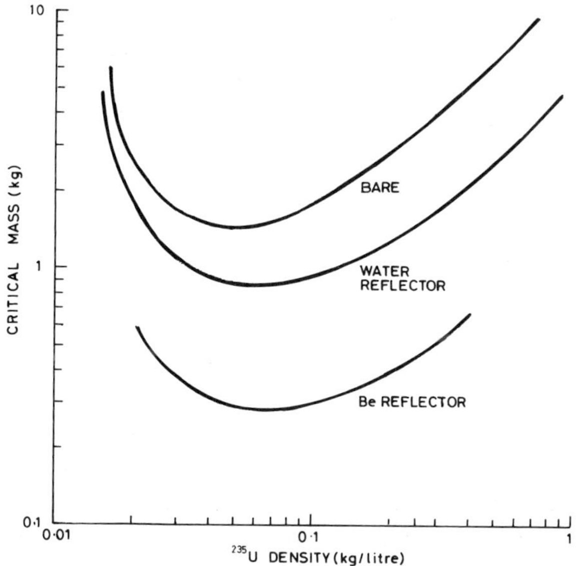

Figure 12.5.3(a) Critical mass versus density for ^{235}U spheres.

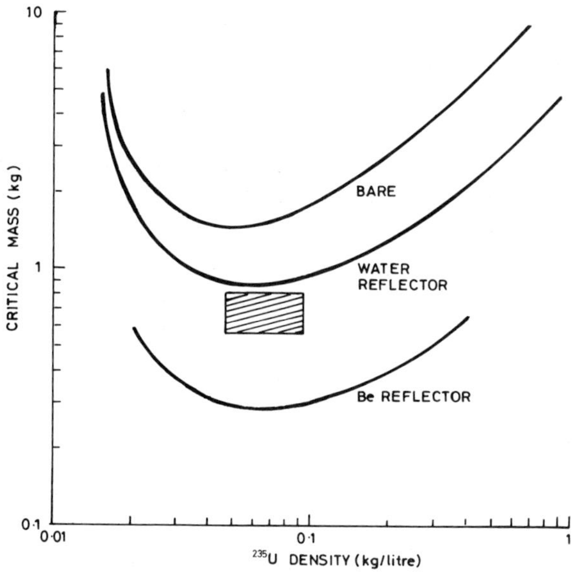

Figure 12.5.3(b) Area of desirable core parameters.

Multi-megawatt requirements in the subsea environment can only be satisfied by nuclear reactor systems. The high energy densities and specific volumes, long mission lifetimes and anaerobic operation provide suitable characteristics for their application as multi-megawatt power sources for total subsea oil production systems. Reactor systems operating at high power levels are more cost-effective and the power conversion systems more efficient. Water is an ideal shield from nuclear radiation and nucleonic considerations indicate that reactor operations in the subsea environment may have improved safety characteristics.[122,125]

The present designs of high power reactor systems for subsea operation are conceptual although considerable experience has been gained with low power systems in space and military programmes. Packaged power systems developed for military and maritime systems (ships, submarines) should be investigated in detail along with the developments in space systems (SNAP). The advanced breeder reactor may also have subsea application in the future and offers the advantage of extended core lifetimes, and therefore longer operational periods before refuelling.

Consideration may need to be given to the concept of distribution of power to peripheral installations from a central nuclear subsea power generation complex. In manned reactor systems this approach may be the only method of providing a cost-effective system by operating at high power levels and easing logistic problems. The possibility of transporting a reactor generation system in a submarine could be investigated, the power being sold by the operating company at the subsea location at some enhanced cost level. Refuelling or replacement of the power system would be standard practice and the oil companies would not need to become involved in reactor operations. Mobile power modules could also be matched to varying field requirements.

12.6 THERMAL SUBSEA POWER SOURCE FOR OFFSHORE OIL PRODUCTION

The concept utilizes the thermal difference between the temperature of oil and seawater to generate power up to a level of 2.5 kWe. The power source is designed to operate, monitor and control systems at individual wellheads.[126]

Crude oil flowing from offshore wells is assumed to have a temperature between 43°C–82°C and ambient seawater 1.7°C–12.8°C. The thermal difference between these two levels is used to create mechanical power using a Rankine cycle, the working fluid being a refrigerant. The gaseous refrigerant expands through a very small turbine which drives a Rice or Lundel type generator. The generator provides A.C. power which is rectified and used to charge Ni/Cd or Ni/Fe batteries to power the systems.

The heat in the oil could provide more energy but the reduction in crude temperature is restricted to 4°C–10°C to avoid changes in viscosity and inhibit paraffin build-up. If production is commingled in a manifold, higher power levels may be extracted because of increased oil flows.

The power supply may have application at remote wellhead sites from the undersea structure if seabottom umbilicals are to be avoided and it may also be capable of producing sufficient power at a manifold centre for emergency life-support if the production flow can be maintained.

12.7 DYNAMIC NON-ATMOSPHERE BREATHING SYSTEMS

12.7.1 IPN Engine

The IPN engine uses isopropyl nitrate monofuel, which decomposes to a gaseous mixture at a minimum pressure of 200 lb/in^2 and a minimum temperature of 200°C, it is an exothermic decomposition so the action is self-sustaining at minimum pressure. The engine will run on the gaseous mixture in the absence of air. Present outputs are of the order of tens of kilowatts and the performance is affected by ambient pressure as the exhaust vents to ambient sea pressure. Power limitations and the undesirability of operating the engine in a standby mode would appear to make it unsuitable for our application.[128]

12.7.2 Closed-Cycle-Diesel

Present development work is centred around the development of a closed-cycle diesel developing tens of kilowatts. Engine exhaust is recirculated via a heat exchanger, scrubber, separator and oxygen injector. Nitrogen is used as a diluent gas for the recycled gases in place of carbon dioxide to increase efficiency and avoid complications when changing between closed-cycle and naturally aspired running conditions. Carbon dioxide, sulphur dioxide and oxides of nitrogen are absorbed in a scrubber using potassium hydroxide. A microprocessor is to be used to control the system operation.[129]

The system will require logistic supply of oxygen and oxygen storage. The closed-cycle diesel will be unsuitable for the large power requirement, but may be applicable to the emergency requirement if the storage system is suitable and starting can be guaranteed.

12.7.3 Stirling Engine

The Stirling engine is an externally heated gas cycle engine which may use any type of matched heat source. The reciprocating engine generally uses helium or hydrogen as a working fluid. The engine exhibits low noise and vibration and low maintenance characteristics and can be desiged for high efficiency (40 – 50%). The incorporation of high temperature sodium heat pipes in the more recent designs has improved engine design and performance.

Phillips are developing a 40 HP engine for underwater use and United Stirling of Sweden are also investigating an underwater system using a gas oil and hydrogen peroxide fuel system. The Stirling engine shows considerable promise but is still under development.[130] Deepwater application will probably involve the use of a radioisotopic heat source or chemical heat sources without gaseous products from the reaction.

12.8 SUMMARY AND CONCLUSION

12.8.1 Normal Operating Power Requirements

It has been estimated that for normal oil field operations a power level in the order of tens of megawatts will be required. Specific requirements will be dependent on field parameters and varying operational demands during the life of the field. The use of gas turbines and diesels for the generation of such power levels above the surface in offshore operations is established. The feasibility of transmitting these levels of power from a shore installation over long distances or from a surface facility through deepwaters raises questions of reliability.

To retain the inherent advantages of a total subsea production facility it would appear fundamental to provide the complex with an autonomous power system, especially if the level of power available allows the product to be pumped ashore or to a surface loading facility. A subsea nuclear reactor powerplant is the only underwater power system capable of generating megawatts of power over several years of operation.

Present generation reactor systems which appear most readily adaptable to the application are the pressurized water reactors or the liquid metal cooled reactors in combination with organic or water Rankine conversion systems. Improvements involving hermetically sealed units, combined rotating units and automatic control promise unattended operations for several years before refuelling and overhaul. Present studies indicate a core life of four years and an overall life of twenty years can be expected. The operation of reactor systems in the underwater environment may exhibit improved safety characteristics.

A requirement has been identified to investigate in detail the present state-of-the-art of reactor systems being developed in the United States for space, military and oceanographic applications, i.e. such programmes as the SNAP, associated with the Atomic Energy Commission and the United States Energy Research and Development Association.

A detailed analysis between reactor systems and air-breathing umbilical-feed systems is required in relation to costs and reliability. A simple guide could perhaps be, that if the cost of an *in-situ* non-atmospheric power plant was less than an air-breathing system, the latter need not be considered.

12.8.2 Emergency Power Requirements

The emergency power supply must be a non-atmospheric power supply. The power must be of sufficient capacity and endurance to provide lighting, life-support, operation of rescue and escape systems and possibly control functions for processing equipment. A major element of the requirement will be for crew heating under survival conditions, other functions will require minimal power.

A level of 10 – 20 kW for 5–7 days is estimated as the emergency requirement. Many of the assessed systems could provide this power level. Lead/Acid batteries could be used if they were 'dry-stored', although Silver/Zinc batteries are preferred as they may be float charged for many years without gassing or electrolyte depletion. The restrictive cycle-life and expense of Ag/Zn would not be critical for this application. Nickel/Cadmium batteries may be suitable if it is possible to ensure that they are fully charged. Lithium thionyl chloride batteries may offer potential advantages.

Fuel cells have high reliability, endurance and efficiency. The considerable developments in this field offer promising potential for their application to undersea activities in the 1980's. The hydrox (H_2/O_2) fuel cell of the membrane type is probably most suitable for the application, because they operate at room temperature and do not experience a start-up delay. Reactant supply would need to be pressurised gas as the 'boil-off' of cryogenic systems make them unsuitable for stand-by applications. Zinc/Oxygen fuel cells may also be of use as they experience no start-up delay and the high cost of anode kits is not restrictive. An added advantage of the hydrox fuel cell is that water and heat are produced as a by-product of the reaction, which could be used to aid crew survival and reduce the emergency electrical demand. The Alsthom and Lockheed cells may also be suitable.

Radioisotopic generator developments based on ^{60}Co Radioisotopic sources could provide sufficient power. The power supply would be continuously available and could be also used to provide a level of redundancy to the primary power source. The cooling system of the generator could provide heating under survival conditions. The disadvantage of continuous operation could be alleviated by utilizing the high radiation fields for the irradiation treatment of waste products from the day to day operations of the complex.

Dynamic systems are unlikely to be suitable, because of the stand-by nature of the application and potential starting problems.

Weight and volume restrictions may limit the use of battery storage, in which case a hybrid of batteries and fuel cells or fuel cells alone would appear most suitable. The selection of a radioisotopic power plant will be determined by the philosophy of operation of the emergency power supply, i.e. should it be continuously available or come on-line when required. Fuel cell or radioisotopic power sources should be integrated at the initial design stage with normal and emergency life-support functions to advantage.

13
SAFETY

The use of a manned underwater complex for subsea oil production will involve the enclosure of highly concentrated process equipment within the structure. The equipment will contain oil and gas under high pressure and leakages could lead to fires or explosions. The crew of the structure are exposed to the potential hazards of this environment on a continuous basis, while on station. They are also exposed to additional hazards during logistic support operations and in the event of emergency evacuation of the complex.

Safety is a difficult concept to visualize, but is generally considered as the protection of life, environment and property, usually by having the ability to withstand or avoid hazards. Safety is not an absolute term, and one should really consider what degree of safety is acceptable in a sociological and economic context. A satisfactory level of safety for a new industrial concept may, however, need to be comparable or lower than those levels encountered in other industries with similar characteristics.

This chapter introduces some basic characteristics of safety analysis and refers to two recent studies that deal specifically with the safety assessment of our concept of a subsea oil production complex. A detailed analysis is not attempted.

13.1 SAFETY CONCEPTS

Safety is normally discussed in terms of risk. Risk is a counterpart to safety in a qualitative sense. Risks can only be evaluated in terms of potential hazards, which in turn bear directly on safety. At the present time, the basis for the management of safety is determined by the analysis of risks, risk analysis being the systematic evaluation and control of undesired events or hazards associated with an activity.

To perform a risk analysis a defined design concept is required. All unwanted events are first identified using past data, theory or experienced subjective views and the causes of these hazards are analysed. The probability of the occurrence of the unwanted event can be determined with a 'fault tree', which is a schematic representation of a combination of causes which may lead to a certain unwanted event.

If the fault probabilities of the basic events are known the overall probability of the unwanted event can be calculated. The consequences from the particular hazard are then analysed using an 'event tree'. The risk associated with the various consequences are also evaluated objectively using all the probabilities and a subjective evaluation of risk made. Possible measures and actions which may reduce the probability of a hazard occurring and list of consequence reducing measures are then assessed in relation to cost.[131]

In summary, the basis for safety management is achieved by the quantification of risks. The risks are identified initially by defining potential hazards and the consequences of the occurrence of the hazard. Probability theory, statistics, operational research and systems analysis are used to evaluate the risks and quantify them in terms of frequency and severity. A programme for risk control is initiated to maximize safety. Finally, technical and economic properties are then modified by sociological considerations, i.e. risk to life, risk to environment, to obtain a subjective evaluation of risk.

13.2 FATAL ACCIDENT FREQUENCY RATE (F.A.F.R.)

The fatal accident frequency rate is a figure that is used as a method for comparing the occupational risk to people in different occupations. It represents the probable loss of life per 10^8 hours of exposure. This time corresponds approximately to the occupational life-time risks for 1000 persons assuming 2000 hours of exposure per year in a 50 year working life.

TABLE 13.2a F.A.F.R. values for industrial occupations.

Industry	*F.A.F.R.*
Chemical Industry	3.5
Steel Industry	8
Fishing	35
Coal Mining	40
Railway Shunters	45
Construction Workers	67
Air Crew	250
Professional Boxers	7000
Jockeys (Flat Racing)	50000

TABLE 13.2b F.A.F.R. values for non-industrial occupations.

Non-industrial Occupations	*F.A.F.R.*
Staying at home	3
Travelling by bus	3
Travelling by train	5
Travelling by car	57
Pedal cycling	96
Travelling by air	240
Moped riding	260
Motor scooter driving	310
Motor cycling	660
Canoeing	1000
Rock climbing	4000

Tables 13.2 a and b from Ref. 135 show some typical F.A.F.R. figures for industrial and non-industrial activities. Offshore field operations, historically find themselves in the higher brackets of occupational risk and at present probably lie in the range 40 - 80.[133] Our concept of subsea production should therefore produce as a maximum a F.A.F.R. comparable with these figures as the activities are similar in many ways.

13.3 SAFETY OF SUBSEA OIL PRODUCTION SYSTEMS

Two recent studies into the safety of our specific concept of subsea oil production system have been undertaken:

1. 'Hazards Associated with Manned Underwater Habitats', Ref. 134.
2. 'Hazard Assessment Subsea Hydrocarbon Production Facility', Ref. 135.

These two reports are directly relevant to the present study, although future system concepts may deviate in configuration from those considered in these reports. The first report is restricted to a qualitative treatment of the hazards, while the second is quantitative and uses the technique of risk analysis described previously. A brief review of each paper is now given, the reader is referred to the original reports for detailed discussion.

13.3.1 'Hazards Associated with Manned Underwater Habitats'

A discussion paper was prepared in May 1978 by the Scottish Offshore Partnership for the Department of Energy into the qualitative assessment of hazards associated with the use of manned underwater structures. The report deals with major potential incidents, i.e. seawater flooding, oil and gas leaks, fire and explosion and hazards of the operating systems. Each area is examined for the potential of faults occurring and the resultant consequences of the faults are considered. The means to deal with these faults, when they occur and the methods of minimizing the hazards by good design are discussed.

Escape methods, maintenance and repair, reliability, systems integrity and operating and emergency procedures are assessed in relation to the potential hazards associated with the operation of the manned underwater habitat.

The report concludes that in planning a complete production facility on the seabed one cannot afford to learn from mistakes. The potential hazards and implications for safety must be recognized and corrected throughout design and building by the use of reliability and safety analysis. Operating and emergency procedures will need to be written before it is too late to correct design errors. The best methods of design from other industries, i.e. oil production, marine, submarine, nuclear and aircraft industries should be selected and used to minimize the hazards. Regulatory authorities should liaise with the designers during the design as well as ahead of it.

13.3.2 'Hazard Assessment Subsea Hydrocarbon Production Facility'

This confidential report was produced for Sir Robert McAlpine & Sons Limited, in February 1979 by the Institute of Offshore Engineering at Heriot-Watt University. The purpose of the study was to produce a criterion upon which the safety of the subsea production facility could be judged in terms of injury or death to operating personnel. The study uses the technique of risk analysis and, within the limitations of the present system design details and failure rate data, provides a quantitative assessment of the underwater complex.

The system is considered as a subsea complex of five cylindrical structures and the normal transfer and emergency escape systems. The installation phase and risk levels to submersible and mothership are not considered. All elements are considered that can affect the safety of the operating personnel, which are assumed to occur in three main areas: the emergency evacuation system, the normal transfer system and a chance of an occurrence in the complex that requires evacuation but the personnel cannot escape.

A 'top-down' or 'fault tree' risk analysis technique was used because of the preliminary nature of the design of the complex and the time limitations on the study. After establishing the undesired event, i.e. fatal injuries, the technique focuses on all equipment failures and human errors which may combine to produce the undesired event together with their probabilites of occurrence. The details of 'fault tree' analysis, data, computation of probabilities and sensitivity analysis are given in the report.

The risk level to the occupants is computed in terms of F.A.F.R. so that a suitable criterion is generated to compare with operating personnel in other industries. A F.A.F.R. value of 25.4 was calculated as the level of risk to which personnel on a normal shift rota would be exposed. The preliminary nature of the study involved many gross assumptions and 25.4 is probably a mean of a probability distribution which extends approximately from 10 to 130.

The level is higher than for many industrial activities, although it has been suggested that a F.A.F.R. of less than 50 is an acceptable risk for an industrial activity.[132] The study concludes that a more detailed investigation is required to reduce the band of error of the F.A.F.R.

13.4 CONCLUSION

The two reports present a preliminary assessment of the potential hazards and risk levels to which the crew of an underwater structure used for oil production may be exposed. In this book it is not possible to expand on this work, but only draw some general conclusions.

Everyday life is full of risks and so are present offshore operations, these risks are more or less appreciated by those involved. A crew member will be primarily concerned with his personal health, an operator primarily with the hazards to his installation and possible liabilities. Regulating authorities will be concerned about a wide range of risks ranging from risk to human life to risks to the environment.

Inherent in the concept of safety is the respect for human life and the environment, with the sociological implications this entails. Safety is therefore not only concerned with the hardware of the activity. Human factors themselves, will in fact also impinge directly on safety through such characteristics as education, training and motivation.

Risk analysis is a very valuable tool in the assessment of safety and presents a more positive attempt to identify hazards before they cause problems and a swing away from the restrospective procedure of accident investigation as the basis for hazard identification. In an attempt to quantify the risks, however, human and sociological factors must not be ignored in the final safety evaluation.

The F.A.F.R. index is a valuable indicator of the risk level of a system in relation to occupational risks in other fields. In 1974 the F.A.F.R. for offshore work was estimated to be in the range 40 - 80. The statistics do not include the recent tragic accident of the *Alexander Kielland* in the Norwegian Sector of the North Sea. This disaster will significantly affect the F.A.F.R. figure for offshore work.

The preliminary study of the subsea complex gave an F.A.F.R. between 10 - 130 (mean probability 25.4), which is comparable and possible even lower than present offshore work. The Burgoyne Committee have recently recommended an improvement in offshore safety; however, the implementation of procedures recommended in the reports could well reduce the levels of risk to the crew of an underwater structure to well below the usually acceptable level for industrial safety, i.e. F.A.F.R. of 50.

REFERENCES

51. *Concepts of Hydrospace Terminals* by M. Yachnis, Consultant Deepwater Structures, Naval Facilities Engineering Command, Washington D.C. American Society of Civil Engineers, July, 1969.
52. *Concept Design for a Manned Underwater Station* by Westinghouse Electric Corporation for United States Naval Facilities Engineering Command, USN CEL. Report CR 67.017, March 1967.
53. *Conceptual Study of a Manned Underwater Station* by General Dynamics Corporation for United States Naval Facilities Engineering Command, USN CEL Report CR 67. 019, April 1967.
54. *Concept for a Manned Underwater Station* by Southwest Research Institute for United States Naval Facilities Engineering Command, USN CEL Report 67.022, February 1967.

55. *Study of One Atmosphere Manned Underwater Structures*, Ocean Systems Operations, North American Rockwell Corporation for the United States Naval Facilities Engineering Command, Report C8-340/020, June 1968.
56. *Technical Information*, Lockheed Petroleum Services, 1979.
57. *Overview of the Use of Subsea One Atmosphere by the Oil Industry Worldwide* by M. Apostolidis, Lockheed Petroleum Services, Society of Naval Architects and Marine Engineers, Eastern Canada Section, April 1978.
58. 'One Atmosphere Wellhead System', *Offshore Services and Technology*, p. 10 (January 1980).
59. *The Role of the One Atmosphere Chamber in Subsea Operations* by B. G. Collip and C. A. Sellars, Shell Development Company, Q.T.C., Houston, 1972.
60. 'The Nuclear Submarine Environment' by R. J. Lambert, D. M. Davies, N. F. Lightfoot and J. E. Morris. *Proceedings the Royal Society of Medicine* **65** (No. 7), 795–800 (September 1972).
61. *Safety and Operational Guide Lines for Undersea Vehicles*, Volumes I, II, III, Marine Technology Society.
62. *Preliminary Results from an Operational 90-Day Manned Test of a Regenerative Life Support System*, N.A.S.A. Langley Research Centre, November 1970.
63. 'The Provision of a Tolerable Environment for Astronauts' by D. Hurden, Bristol Siddely Engine Division, *Journal Institute of Heating and Ventilation*, pp. 71–78 (June 1966).
64. *Scientists in the Sea Tektite 2*, United States Department of the Interior, August 1971.
65. 'Space Laboratory Environmental Control/Life Support System — Design Safety' by Richard Gartner, Dornier Systems GmbH. Germany, *The Journal of the Astronautical Sciences.*
66. *Comparative Evaluation of Environmental Control and Life Support Systems for Space Shuttle Orbiter* by D. M. Morris, N.A.S.A. Manned Spacecraft Centre, American Society of Mechanical Engineers, August 1972.
67. *Space Construction Base Support Requirements for Environmental Control and Life Support Systems* by R. J. Thiele, T. C. Secord and G. L. Murphy, American Society of Mechanical Engineers, July 1977.
68. *Space Materials Handbook* 3rd edition. SP. 3006, N.A.S.A., 1969.
69. *Bio-astronautics Data Book* SP. 3051, N.A.S.A., 1964.

70. 'Closed Environmental Problems in Nuclear Submarines' by D. M. Davies, *Ann-Occup. Hyg.* **17**, pp. 239–244 (1975).

71. *Physiological and Hygienic Considerations determining Optimum Atmospheres for Spacecraft Cabins* by N. A. Agadzahanyan, Consultant Nureau, Plenum Publishing Corporation, 1969.

72. 'Sealed Environments in Relation to Health and Disease' by Captain J. Shuttle, *U.S.N. Archives of Environmental Health* 8, pp. 438–452 (March 1964).

73. 'Carbon Dioxide and Nuclear Submarines' by M. G. Harper and D. M. Davies, *Journal Royal Naval Scientific Service* **29** (No. 2).

74. *The Effects on Man of Continuous Exposure to 0.5% Carbon Dioxide* by D. M. Davies, D. Smith, D. R. Leitch and J. E. Morris, Institute of Naval Medicine, Hants, July 1976.

75. 'The Application of Threshold Limit Values for Carbon Monoxide Under Conditions of Continuous Exposure' by D. M. Davies, *Ann. Occup. Hyg.* **18**, pp. 21–27 (1975).

76. *Forecast of Human Factor Technology and Requirements for Advanced Aero-Hydro Space Systems* by H. E. Price and J. F. Parker, Jr. Biotechnology Inc. for the Office of Naval Research, March 1971.

77. *Subsea Hydrocarbon Production Normobaric Habitat System-Summary Report on Medical and Psychological Aspects* by Dr. D. M. Davies for Sir Robert McAlpine, January 1979.

78. *Skylab Experiment M487 Habitability/Crew Quarters* by C. C. Johnson, N.A.S.A., Lyndon B Johnson Space Centre.

79. *Habitability Issues in Long Duration Undersea and Space Missions* by J. F. Parker and M. G. Every, Biotechnology Inc. for Office of Naval Research, July 1972.

80. *Advanced Biotechnology for Underwater Operations*, Charles F. Gell Naval Submarine Medical Research Laboratory, Naval Submarine Medical Centre, New London, Groton, Conn., February 1971.

81. *Studies of Social Group Dynamics Under Isolated Conditions*, The George Washington University Medical Centre for N.A.S.A., December 1974.

82. *Psychological Problems of Prolonged Marine Submergence* by Benjamin B. Weybrew, United States Naval Medical Centre, Groton, Conn., October 1971.

83. *Submarine Crew Effectiveness During Submerged Missions of Sixty or More Days Duration*, Benjamin Weybrew, Naval Submarine Medical Centre, Groton, Conn., October 1971.

84. 'Psycho-Sociological Problems of Small Isolated Groups Working Under Extreme Conditions' by R. Angiboust, F. Maline, ed. in *Life Sciences Research and Lunar Medicine*, Pergamon Press, New York, 1967.
85. 'The Effects of Confinement on Long Duration Manned Space Flights', Proceedings N.A.S.A. Symposium, Office of Manned Space Flight, Washington D.C., November 1966.
86. 'The Composition of Groups. A Review of the Literature' by W. Haythorn. *Acta. Psychologica* 28, pp. 27–128 (1968).
87. *International Review of Manned Submersibles and Habitats* by Joseph Vadus, N.O.A.A. United States Department of Commerce, 1975.
88. *International Status and Utilisation of Undersea Vehicles 1976* by Joseph Vadus, N.O.A.A. United States Department of Commerce, 1976.
89. 'Engineering Aspects of Manned and Remotely Controlled Vehicles' by R. Frank Busby, in *Sea Floor Development: Moving into Deep Water*, Royal Society, London, June 1977.
90. *Manned Submersibles* by F. Frank Busby for the Office of the Oceanographer of the Navy, 1976.
91. Technical Information from: (a) Vickers Slingsby — L5; (b) Bruker Meerestechnick GmbH — Mermaid; (c) Perry Oceanics — P.C. 16; (d) Kockums — U.R.F.; (e) International Hydrodynamics — Taurus; (f) Gabler Marine (U.K.) — Tours Submarine/Rescue Capsule.
92. 'Manned Submersibles of Today and Tomorrow', *Offshore Engineer*, pp. 12, 13 (January 1980).
93. 'The Use of Submersibles in the Offshore Industries' by J. N. Mansfield and T. Howard-Jones, The Society for Underwater Technology at Oceanology International, 1978.
94. 'The United States Navy's Submarine Rescue Vehicle', *The Naval Architect*, pp. 12, 13 (January 1973).
95. 'D.S.R.V.'s Now Operational, Rapidly Deployed', *Sea Technology*, p. 27 (November 1979).
96. 'D.S.R.V. Mates with "Stricken" Submarine, Rescue Crew', *Sea Technology*, (July 1979).
97. 'The Role of the Mini-Submarine in the Offshore Field' Gabler Marine in *Petroleum Times*, pp. 18–21 (April 1977).
98. 'Operational Experience with Atmospheric Diving Suits' by T. C. Earls and D. S. Fridge, Oceaneering International and J. P. Balch, DHB Construction, O.T.C., Houston, May 1979.
99. 'Spider Ready to Dive in the North Sea', *Offshore Engineer*, p. 28 (January 1980).

100. *A Pilot Study on Diver Replacement* by R. L. Allwood and A. J. Fyfe, Heriot-Watt University, Marine Technology Centre, February 1979.
101. 'Escape from Submarines: A Short Historical Review of Policy and Equipment in the Royal Navy' by H. J. Tabb. R.C.N.C., Royal Institute of Naval Architects, Spring Meeting 1974.
102. 'The Submarine that Saves its Crew' by B. L. Friedman United States Office of Naval Research, *Naval Research Reviews*, p. 23 (1970).
103. *International Workshop on Escape and Survival from Submarines*, Naval Submarine Medical Research Laboratory, Groton, Conn., Ed. C. F. Gell and J. W. Parker, October 1974.
104. 'Development of the Royal Navy Submarine Escape System' by K. E. Taylor, *Engineering in Medicine*, pp. 104–110 (October 1972).
105. 'Submarine Escape from 600 ft Using Rapid Compression and Buoyant Ascent' by D. H. Elliott, British Navy Staff, Washington D.C., O.T.C., Houston, 1971.
106. *Human Factors Evaluation of Submarine Escape* by B. Ryack and G. Walters, United States Naval Submarine Medical Centre, Groton, Conn., October 1970.
107. 'Submarine Escape Capsules' by Peter Kruger, Ingenieurkentor of Lubek, Oceanology International 80, Brighton 1980.
108. Personal Communication. J. F. Collins, Deep Submergence Systems Project, Department of the Navy, Naval Sea Systems Command, Washington D.C., February 1981.
109. 'Floating Powerplant to Support Submerged Offshore Operations' by R. N. Edwards and J. E. Zupenick, General Electric Company, O.T.C., 1969.
110. *Total Subsea Production System: Subseamac One*, Sir Robert McAlpine and Sons Limited, circa 1977.
111. *A Survey of the Future for Electrochemical Power Sources Undersea* by G. T. Rogers, K. J. Taylor, P. T. Moseley and A. D. Turner, Materials Development Division. A.E.R.E. Harwell, February 1978.
112. *Energy Systems of Extended Endurance in 1 – 1000 kW Range for Undersea Applications*, National Academy of Sciences, U.S.A., 1968.
113. *Non-Nuclear Powerplants for Underwater Propulsion* by Eleanor McNair, Society of Underwater Technology, February 1975.

114. *Subsea Power Generation — A Feasibility Analysis* by R. V. Thompson, University of Newcastle, 1978.
115. *Power and Heat Production Systems* by B. J. Tharpe and T. S. May, General Electric Company, Arctic Institute of North America, 1971.
116. *Fuel Cells — A Survey* N.A.S.A., SP5115, 1973.
117. 'Hydrogen-Oxygen Alkaline Fuel Cells' by P. Bolan and B. Gitlow, United Technologies Corporation, National Fuel Cell Seminar, June 1977.
118. 'Fuel Cell Powers Deep Test Dive', *Petroleum Engineer International*, (January 1980).
119. *Status of 40 kW and 4.8 MW Fuel Cell Programs*, L. H. Handley, United Technologies Corporation, July, 1978.
120. *Radioisotopic Thermal Energy Sources for Space Station Life Support Systems* by R. V. Elms, Jr. and H. H. Greenfield, Lockheed Missiles and Space Company, 1970.
121. *Handbook of Ocean and Underwater Engineering* by J. Myers, C. H. Holm and R. F. McAllister, North American Rockwell Corporation.
122. *An Evaluation of Undersea Nuclear Power Systems* by Walter H. D'Ardenne, Nuclear Engineering Department, The Pennsylvania State University, circa 1970.
123. 'Subsea Nuclear Power Systems for Future Offshore Production Operations' by A. B. Martin, Atomics International Division, Rockwell International, 11th. I.E.C.E.C., 1977.
124. *Deep Sea Production System*, Vol. 4, *Power Supplies*, by Rolls Royce Limited and British Insulated Calender's Cables Limited, 1978.
125. 'Reactor Powerplant for Undersea Application' by R. I. Brasier and C. B. Mills, University of California, Los Alamos Scientific Laboratories, New Mexico, *Nuclear Technology* 22 (May 1974).
126. *A Subsea Power Source for Offshore Oil Production* by W. Dixon and H. Sheets, University of Rhode Island, I.E.E.E. Ocean, 1975.
127. *A Survey of Dynamic Power Sources for Underwater Applications* by J. A. S. Guthrie, J. G. Davies and M. H. Moore, Underwater Engineering Group of CIRIA Report No. U.R. 4, December 1973.
128. *Power Pack for Underwater Application* by J. Overy and E. J. Rayner, Plessey — Systems Technology. No. 12, April 1971.
129. *Development of an Underwater Power Generation System*

by R. V. Thompson and M. Hargreaves, University of Newcastle, November 1978.

130. *Stirling Engine Design Studies of an Underwater Power System and Total Energy System* by H. A. Jaspers and F. K. du Pre, Phillips Laboratories, New York, circa 1973.

131. *Risk Analysis Offshore* by Bjorn Mykatun and *Risk Analysis* by Odd Treit, Veritas, December 1978.

132. *A Preliminary Approach to Risk Analysis for Deep Diving Systems* by G. Evensen, Det Norske Veritas, Report No. 75-284.

133. *Hazards of Offshore Operations, Appraisal and Control* by H. F. Grorud and C. Boc, Det Norske Veritas, Automation in Offshore Field Operation, North Holland Publishing Company.

134. *Hazards Associated with Manned Underwater Habitats* by D. R. McKellar, Scottish Offshore Partnership for Department of Energy, May 1978.

135. *Hazard Assessment Subsea Hydrocarbon Production Facility* by A. Lyons, Institute of Offshore Engineering for Sir Robert McAlpine and Son Limited.

136. *The Role of the Department of Energy in the Safe Development of U.K. Offshore Oil and Gas Prospects* by H. R. George, Department of Energy.

137. 'Safety Aspects Related to Hydrocarbon Production and Processing Offshore Plants' by O. Haun, H. R. Hansen, J. Nylund and O. A. Hofnor, Det Norske Veritas, O.T.C., Houston, May 1979.

138. *Safety Characteristics of Lockheed's Subsea Production System* by G. H. Fahlman, Lockheed Petroleum Services, O.T.C., Houston, May 1974.

SECTION C

Systems Analysis, Related Technology and Future Subsea Configurations

14
GENERAL INTRODUCTION

A diagrammatic analysis of the topics investigated in Section B has been performed. System parameters are correlated with operational requirements and where possible the state of development of the various options is identified. Each topic is then individually analysed in relation to all other topics to indicate the inter-relation between the systems or areas of concern. A general analysis is performed, because the tentative nature of the study will not allow specific definition of the system configuration.

The analysis of system parameters is in essence a simplified presentation of the options identified in Section B. The interrelationship between topics has been obtained by considering how each individual topic affects all the other components of the total system and how it in turn is affected by the component systems. Some interacting parameters identified for each topic are shown. The method of analysis does involve some inherent duplication, but each diagram will represent the topic considered. To identify all the interactions and optimize the overall system design, a full matrix analysis of a total system configuration will be required, when a full system concept has been finalized.

Other technology related to the use of manned underwater structures for deepwater offshore oil production is considered. The use of surface piercing structures for product export and possible developments in product transportation systems are reviewed. Present research work on pressure hull construction, fabrication, processing equipment and power supplies is identified. General ocean technology is discussed in relation to the use of large manned underwater structures.

On the basis of possible developments which have been identified in the earlier work, an attempt is made to configure some alternative production systems for deepwater offshore exploitation.

The configurations are developed progressively from hybrid subsea systems, relying on a relatively high level of surface support towards a more truly subsea facility. The final configuration considers the idealized situation, where several offshore fields are developed as an integrated system. A speculative approach is used to generate various scenarios to indicate the direction in which future deepwater offshore production systems may develop.

A general summary assesses the configurations and associated sub-systems and identifies technological limitations to the subsea approach. Future research and development requirements are specified.

15
SYSTEMS ANALYSIS OF EACH TOPIC

The analysis of each topic provides a summary presentation of the information gathered in Section B. The diagrams (Figs 15.1–15.9) are self-explanatory and the descriptions given only provide general observations on each topic.

15.1 LIFE SUPPORT SYSTEMS AND ENVIRONMENTAL CONTROL

The analysis identifies three main areas; normal and emergency operating conditions and possible future developments in life support systems. There are no technological feasibility problems with life support systems, but the future developments indicated should be investigated. A permanent manned underwater station that cannot be vented to the atmosphere intermittently may exhibit a problem due to the long-term build-up of trace contaminants.

15.2 PHYSIOLOGICAL IMPLICATIONS OF ATMOSPHERE CONTROL

Recommended levels for major atmosphere constituents based on nuclear submarine and space systems are shown. Maximum permissible concentrations for trace contaminants for the specific operating environment will need to be established. The effect of a two week crew rotation schedule on physiological parameters may need to be investigated, as the intermittent exposures may alleviate physiological effects. Medical monitoring of on-going crew will be required to inhibit the transfer of pathogenic organisms to the enclosed environment.

15.3 HUMAN FACTORS

The importance of habitability and system design on the satisfaction of human requirements is shown. Human factors will also be related to environmental control and life support functions through their effect on physiological functions. Psycho-social factors will also be influenced by human factors.

15.4 PSYCHOLOGICAL ASPECTS

These aspects are analysed in relation to crew composition and environmental factors, human factors and life support systems will also interact directly with psychological aspects.

15.5 RESCUE AND TRANSPORT

The various combinations of support vessel, support submarine and bell/submersible operations are considered. An autonomous submarine acting as the base for submersible operations would appear most appropriate. A stabilized submarine could serve as a surface or seabed stand-by vessel and carry out transfer operations subsea, returning to the surface when conditions permit, to recharge batteries or return to base. Developments and limitations in dry hatch transfer and docking arrangements should be investigated in detail. The feasibility of operating an autonomous submarine support vessel with a submersible for the requirements should be assessed.

15.6 ESCAPE SYSTEMS

Escape systems can be divided into individual or group escape. The one-atmosphere buoyant ascent capsule carrying a group of men appears to be the most appropriate system for our requirement. The individual escape suit has very limited application, only to 180 m, although the adaptation of JIM type suits for individual one atmosphere buoyant ascents could be investigated as a secondary escape method.

15.7 NORMAL POWER

The development of deepwater fields is likely to be at a distance from the shore, the use of transmission cables for power supply may be impractical. Surface generation and supply by umbilical may be unreliable. A subsea nuclear reactor is the only feasible means of generating megawatts of power underwater at the field location. Developments in underwater nuclear power plants should be investigated.

15.8 EMERGENCY POWER

Storage problems with batteries and the unreliability of dynamic systems, after long stand-by periods, appear to make them unsuitable for this application. Radioisotopic generators are recommended for systems employing umbilical feed primary power systems, and fuel cells with non-umbilical primary power supplies. Development in radioisotopic and fuel cell powerplants should be investigated for this application.

15.9 SAFETY

Safety considerations will affect all operational systems, crew exposure to hazards within the station while on duty will be continuous. The crew will also be exposed to additional hazards during crew transfer operations or escape and rescue procedures. Safety will indirectly affect psychological aspects.

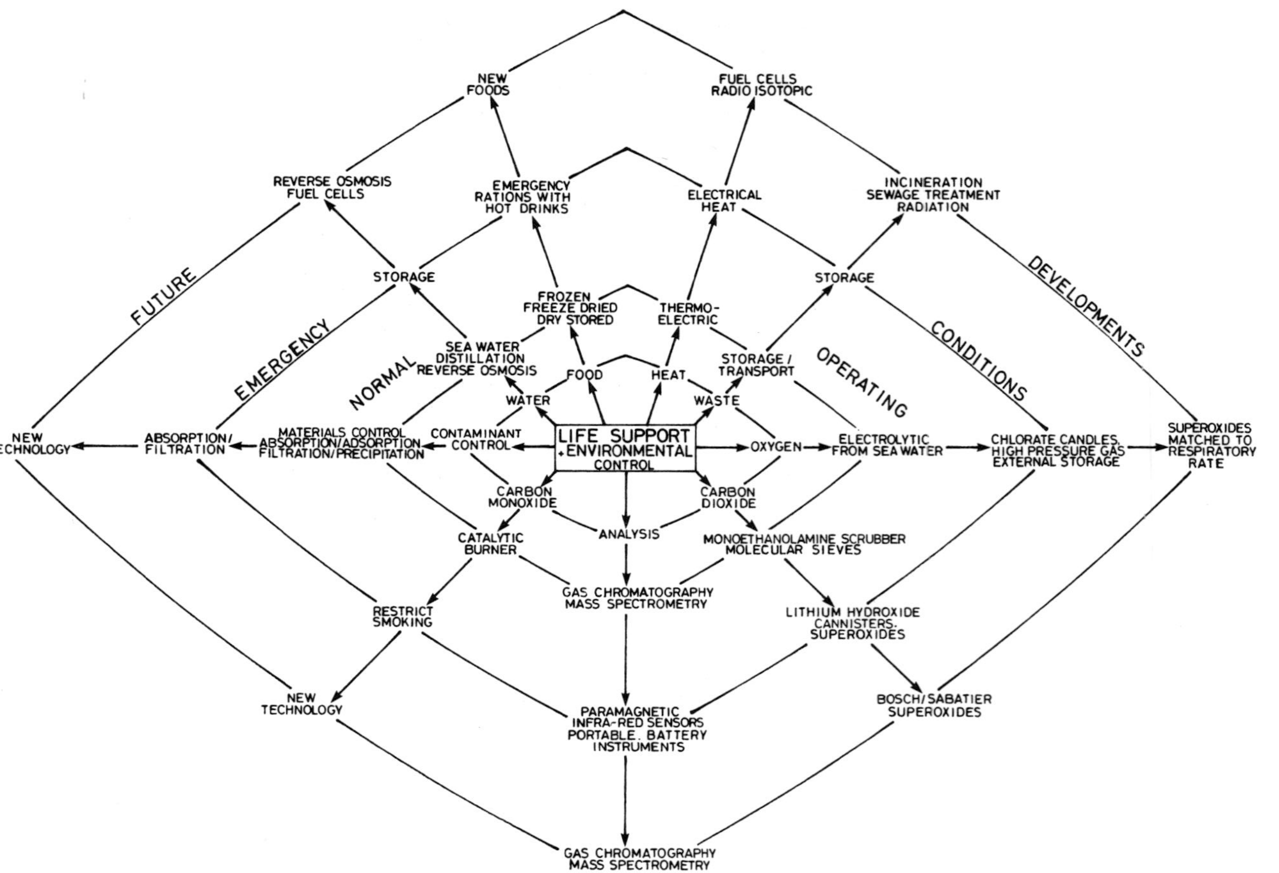

Figure 15.1 Life support and environmental control.

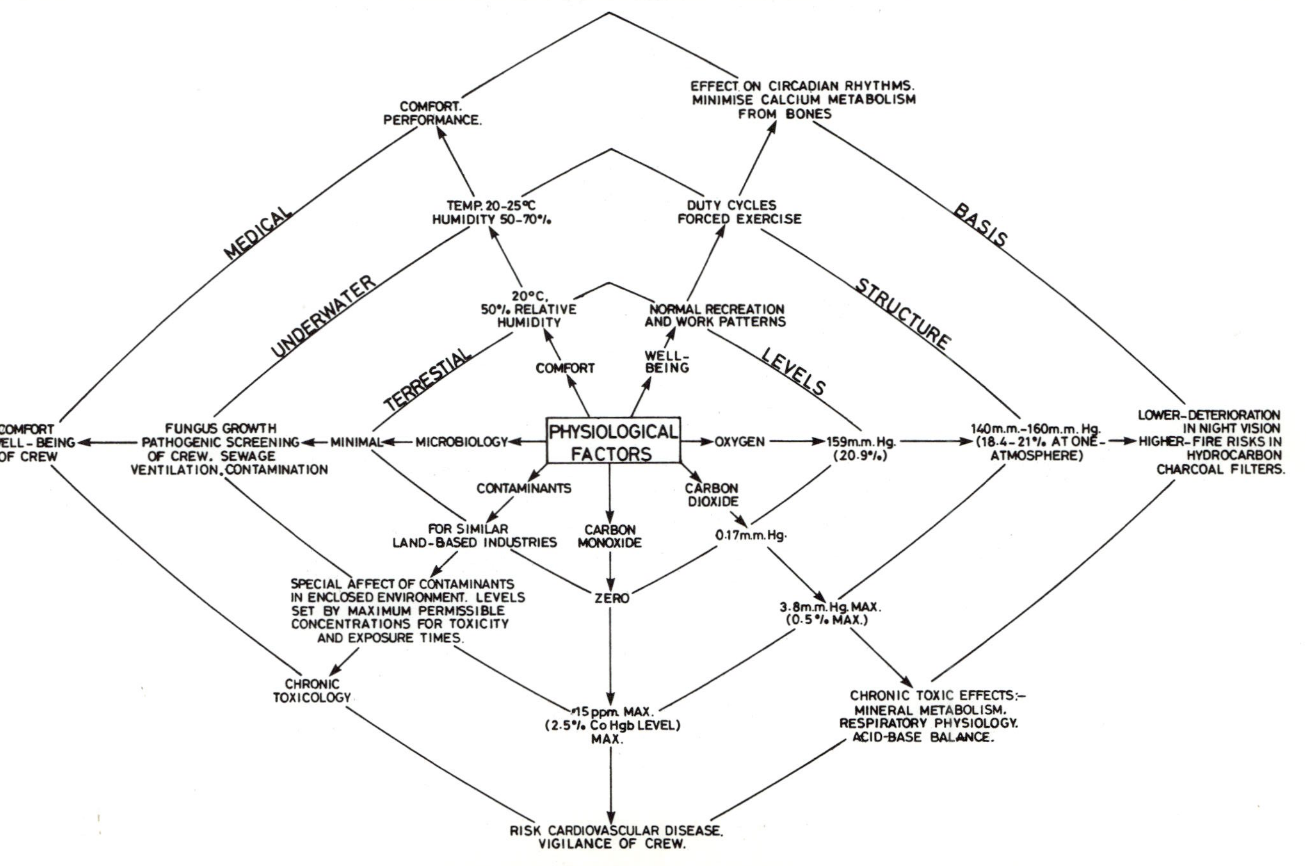

Figure 15.2 Physiological factors.

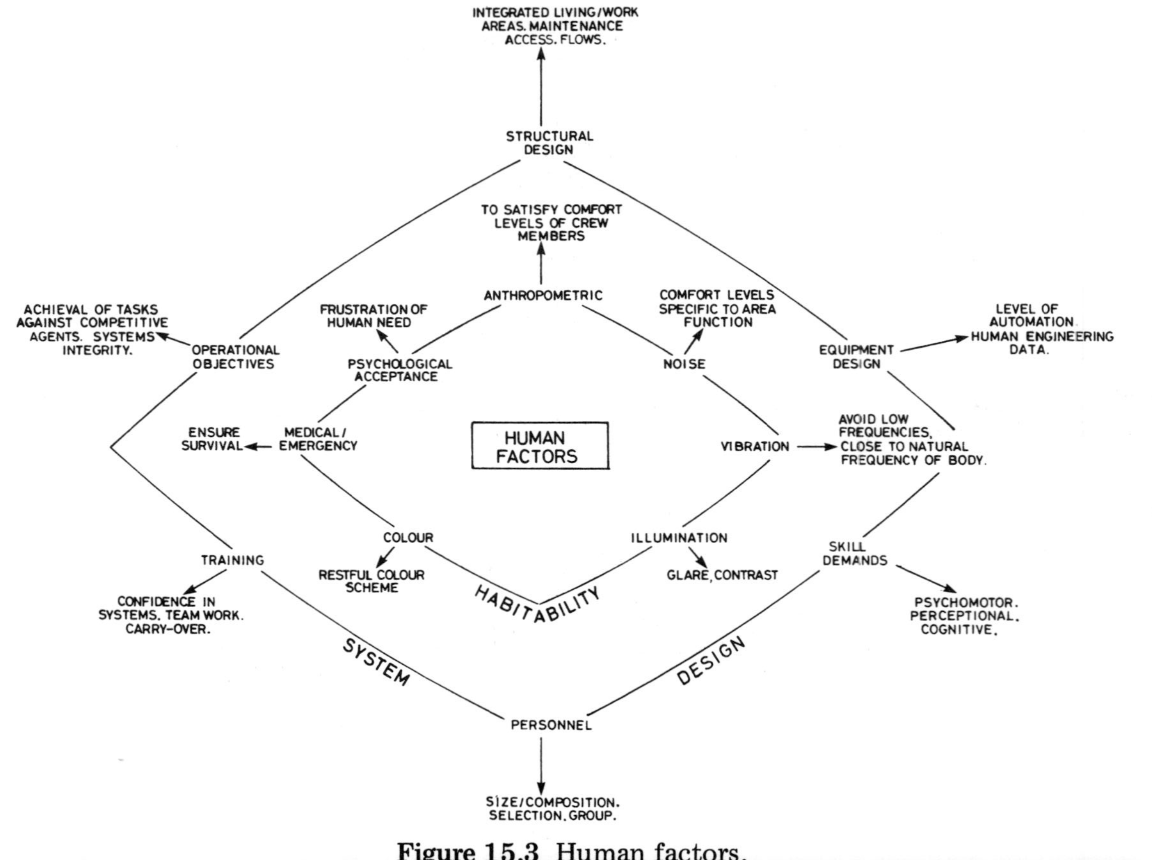

Figure 15.3 Human factors.

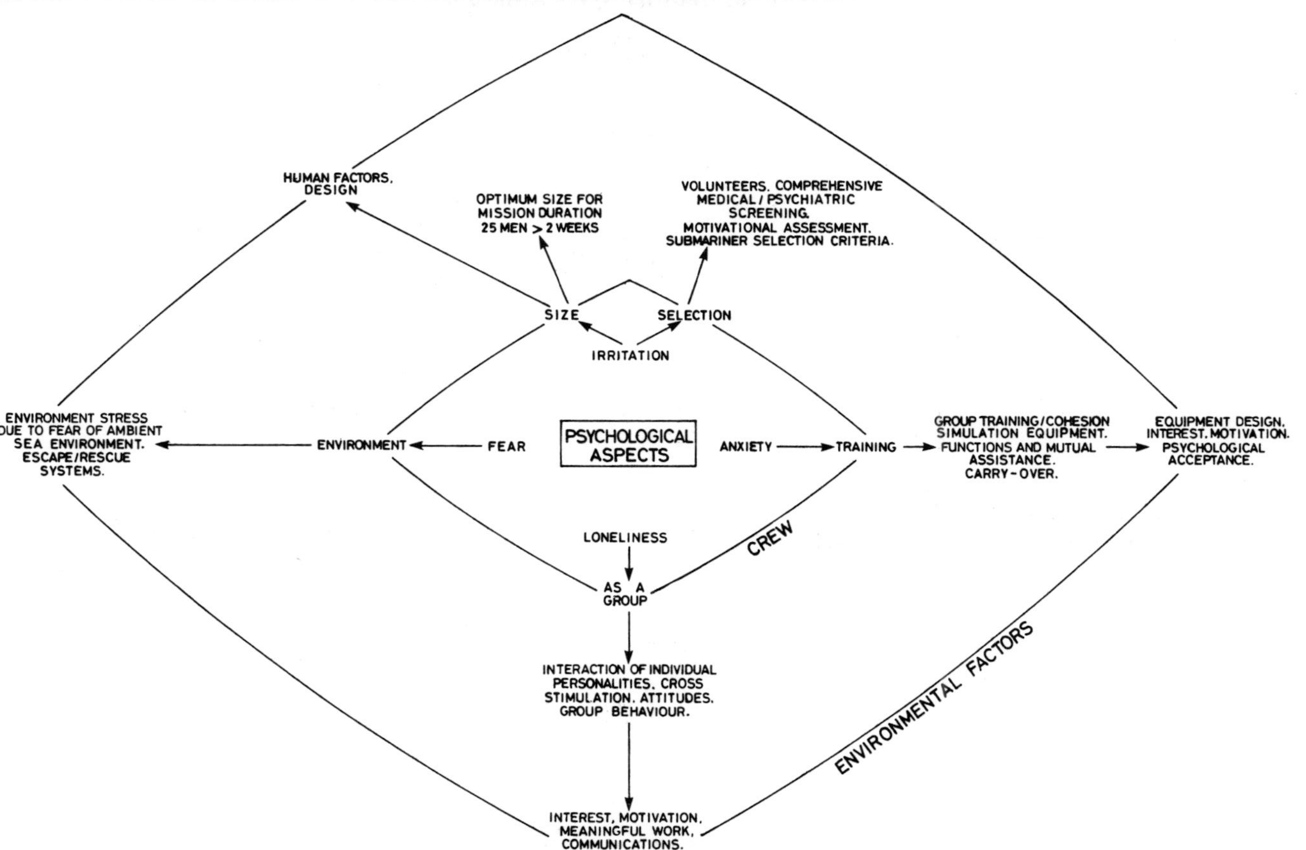

Figure 15.4 Psychological aspects.

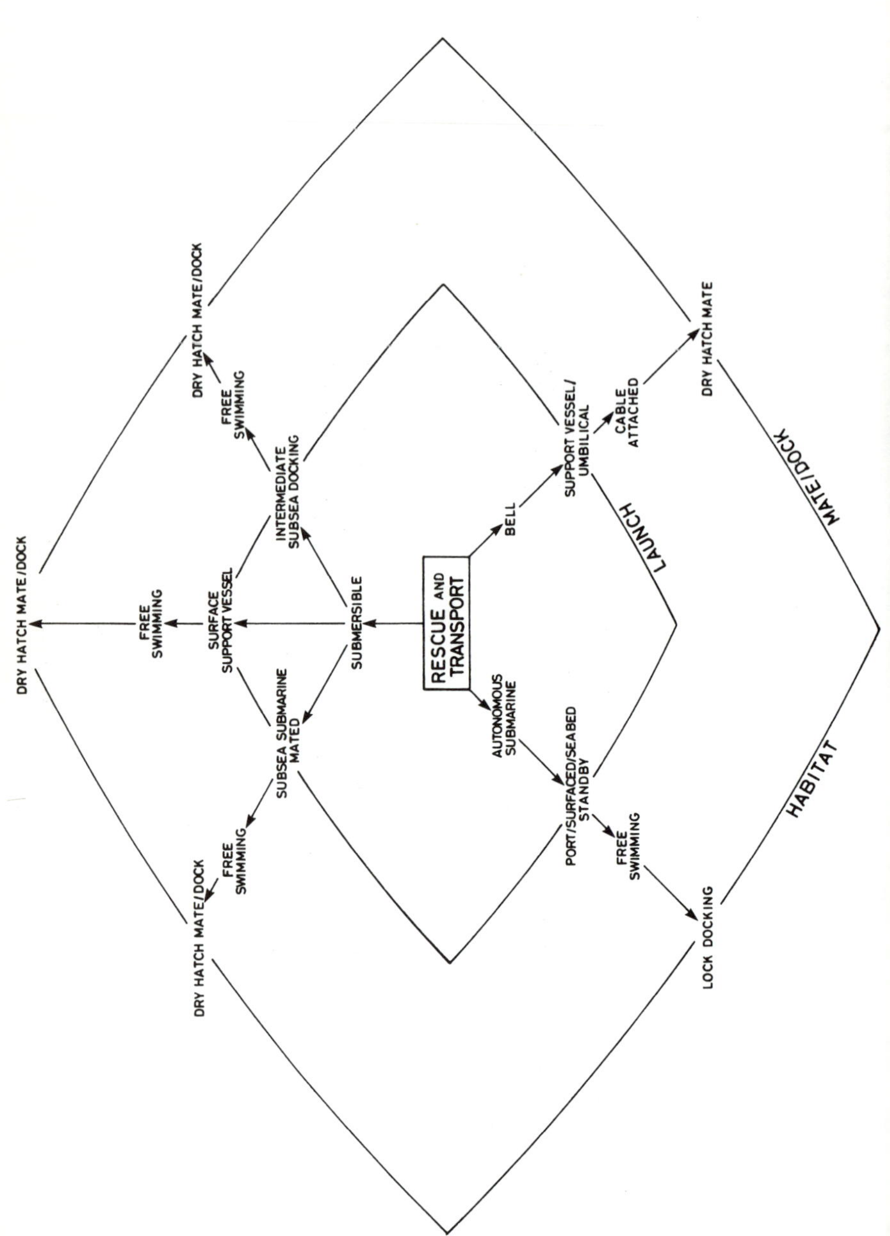
RESCUE AND TRANSPORT
SUBMERSIBLE
SURFACE SUPPORT VESSEL
FREE SWIMMING
DRY HATCH MATE/DOCK
INTERMEDIATE SUBSEA DOCKING
FREE SWIMMING
DRY HATCH MATE/DOCK
SUBSEA SUBMARINE MATED
FREE SWIMMING
DRY HATCH MATE/DOCK
BELL
SUPPORT VESSEL/ UMBILICAL
CABLE ATTACHED
DRY HATCH MATE
AUTONOMOUS SUBMARINE
PORT/SURFACED/SEABED STANDBY
FREE SWIMMING
LOCK DOCKING
LAUNCH
MATE/DOCK
HABITAT

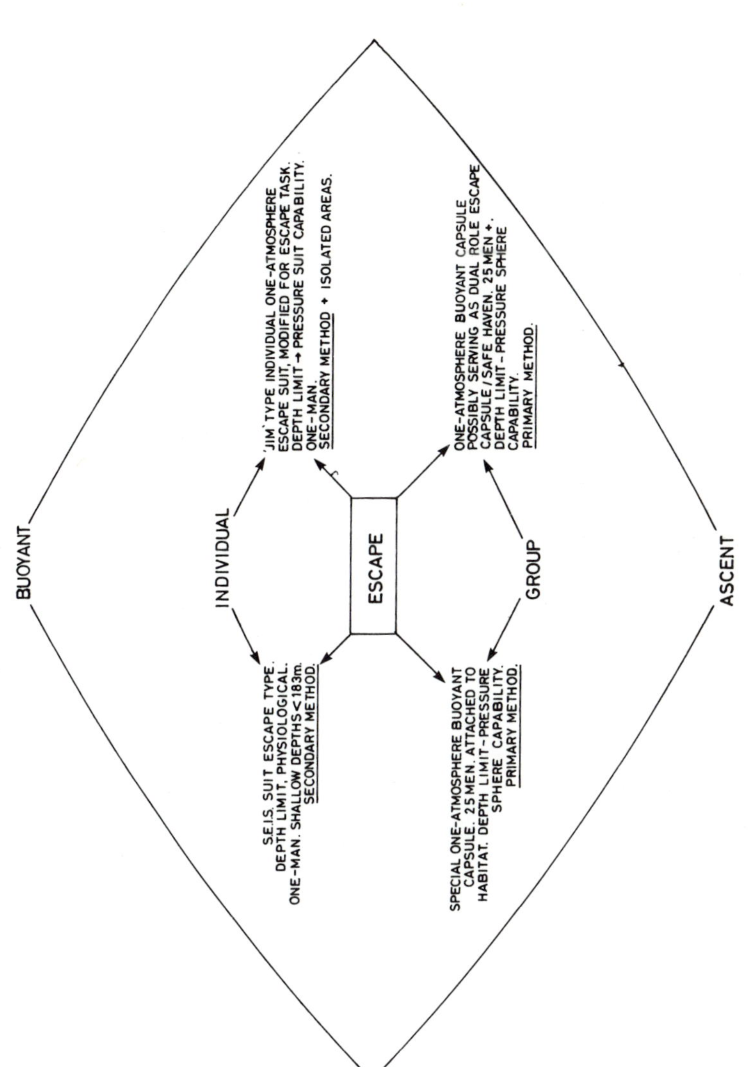

Figure 15.6 Escape.

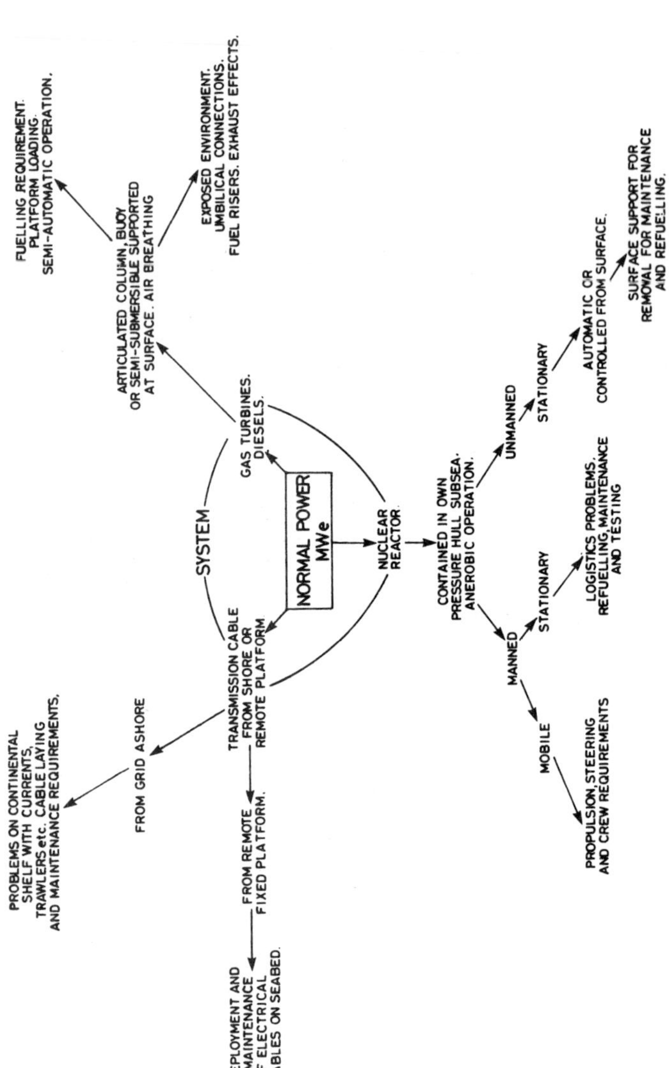

Figure 15.7 Normal power.

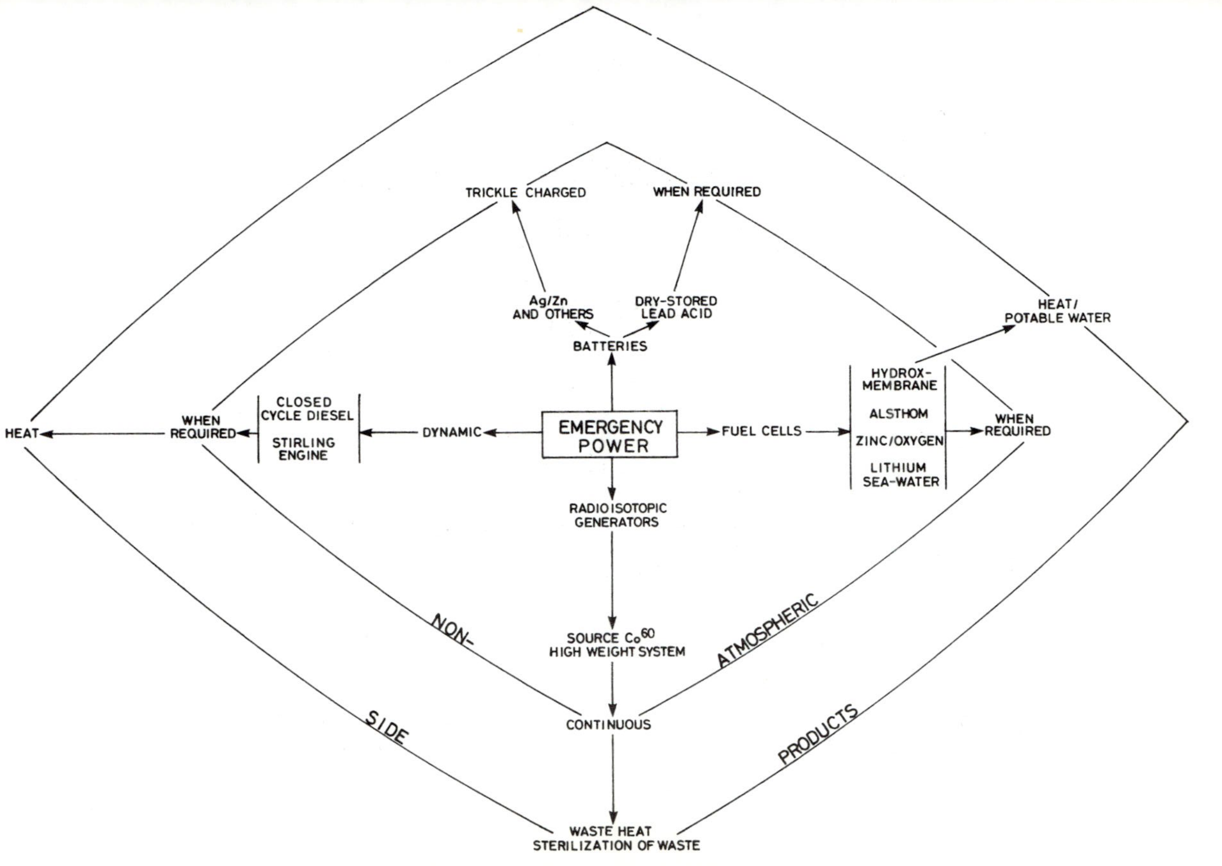

Figure 15.8 Emergency power.

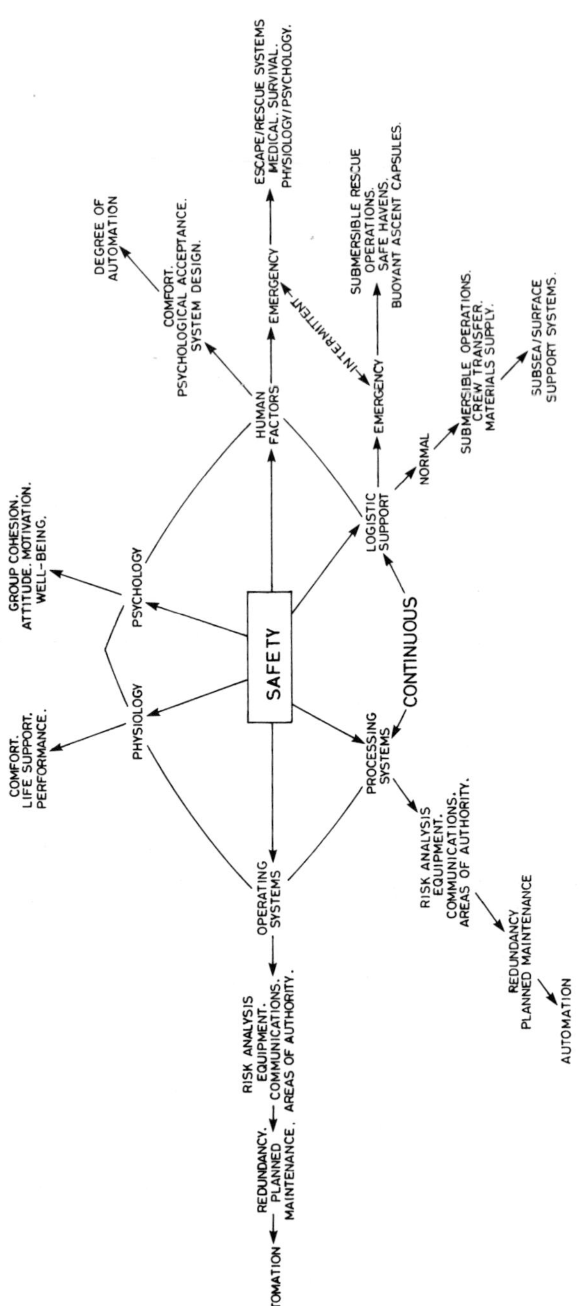

Figure 15.9 Safety.

16

INTERACTION BETWEEN SYSTEMS

The topics of interest are analysed for the interaction that occurs between them. The central topic for each analysis is investigated for its affect on the associated systems, additionally the affect of the associated systems on the chosen topic is considered. Brief descriptions of interacting functions are noted for clarification. This method of analysis does involve some duplication in the presentation, but allows each diagram to represent each topic fairly concisely.

Two additional topics have been considered in the total analysis, these are 'Manning Level' and 'Processing Equipment'. The two functions will bear directly on many of the topics considered. All interacting functions are not described, a full investigation of the interactions will require a matrix analysis of the total system configuration to optimize the system design. The design state does not allow this at present, and time was not available to undertake this type of analysis on the systems postulated later in this section.

The presentation is again self-explanatory and no attempt is made at this stage to specify system requirements. The aim is to provide a qualitative appreciation of the complex interactions which will occur between sub-systems and areas of interest. Further detailed analysis will be possible once a system configuration has been finalized (Figs 16.1 - 16.12).

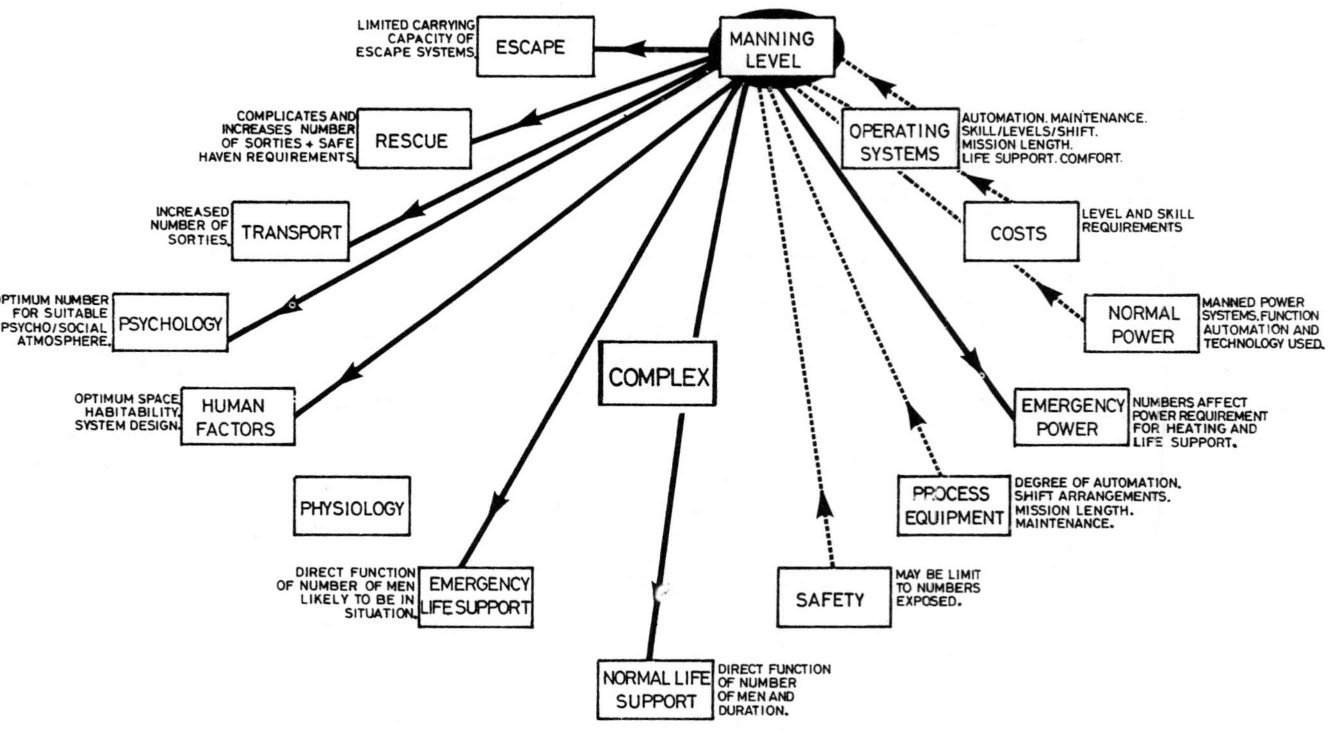

Figure 16.1 Manning level.

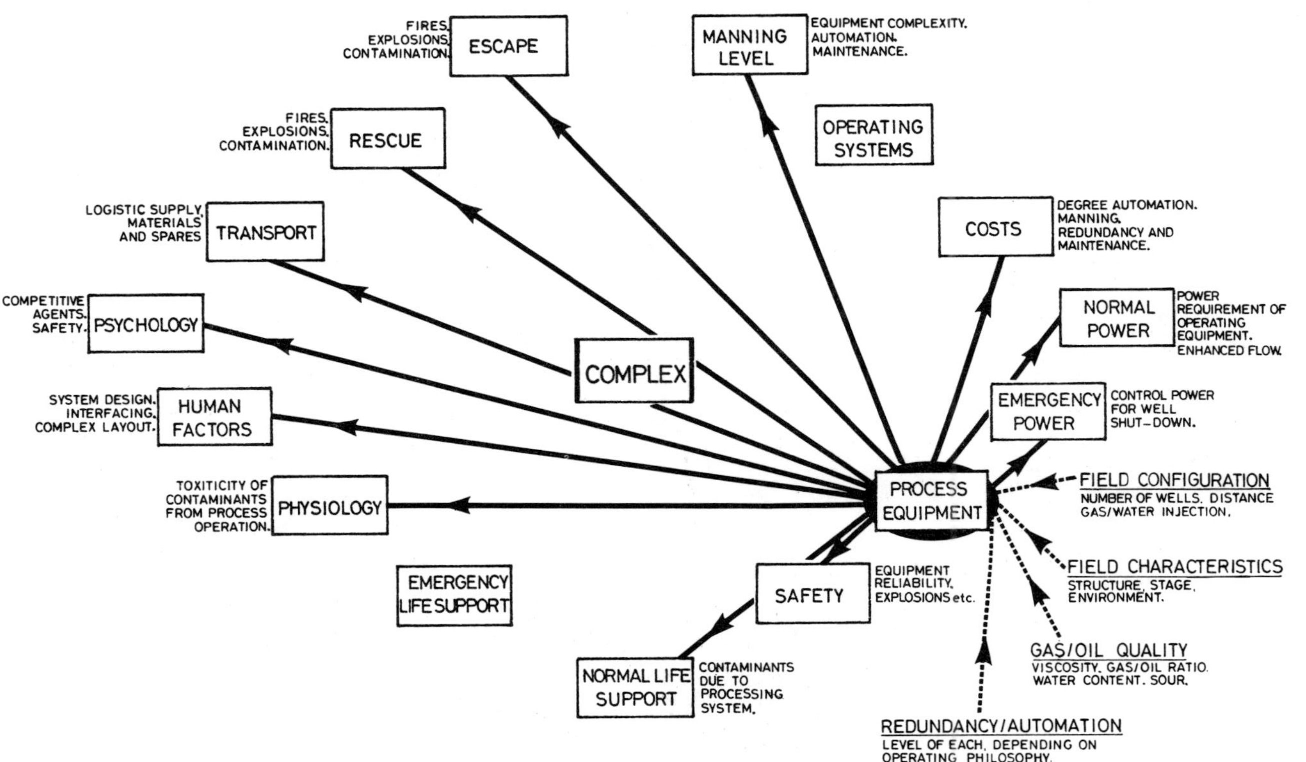

Figure 16.2 Process equipment.

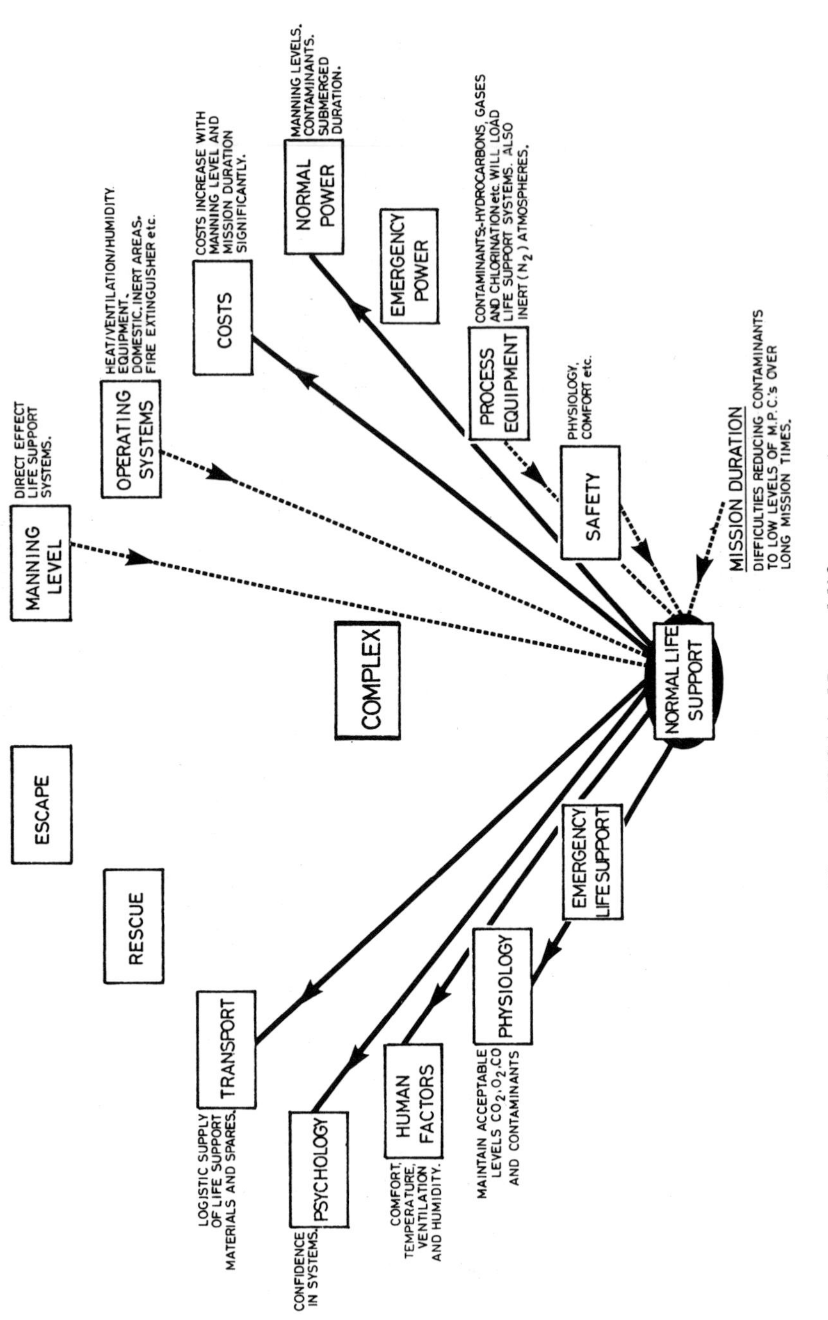

Figure 16.3(a) Normal life support.

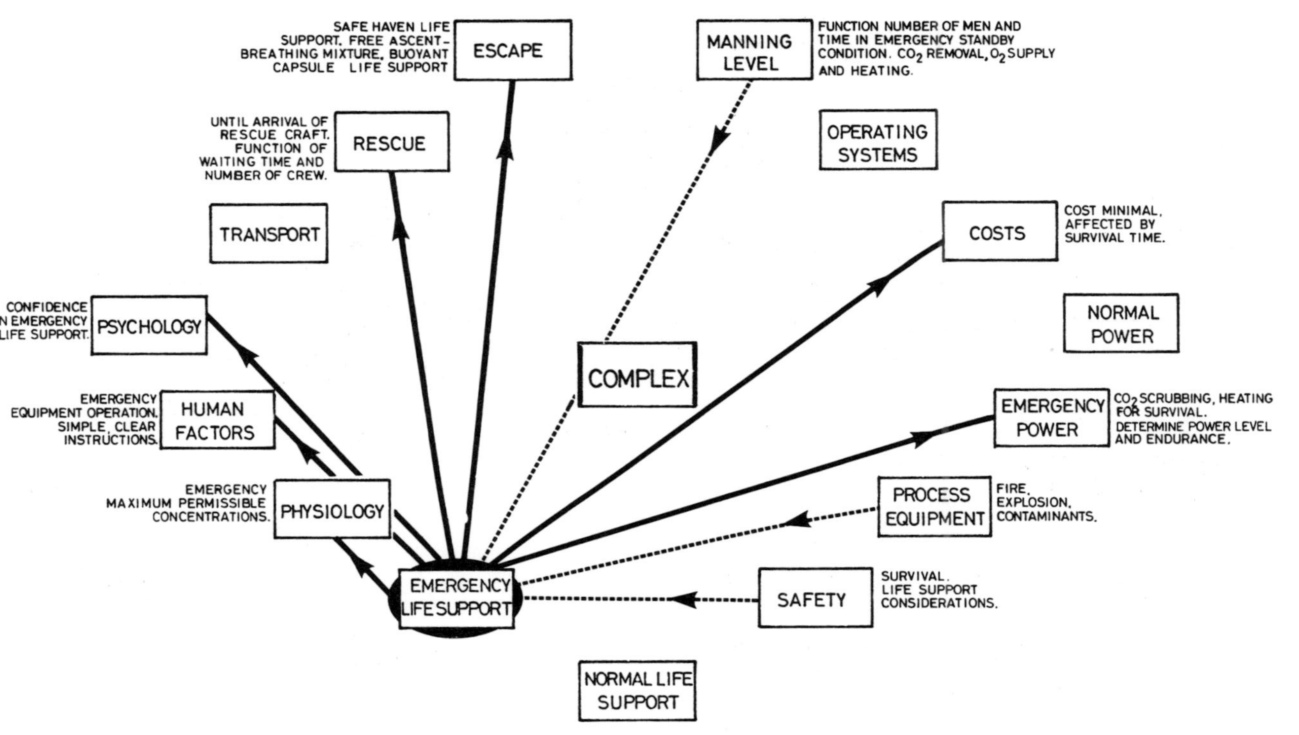

Figure 16.3(b) Emergency life support.

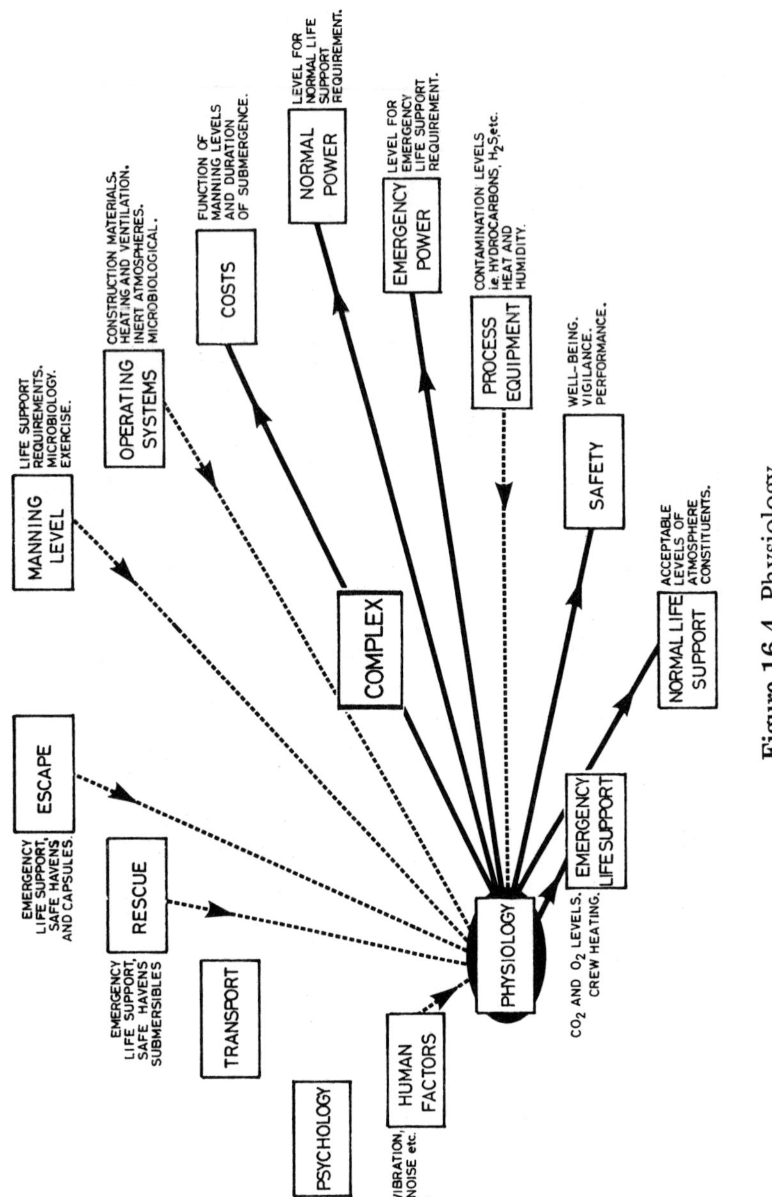

Figure 16.4 Physiology.

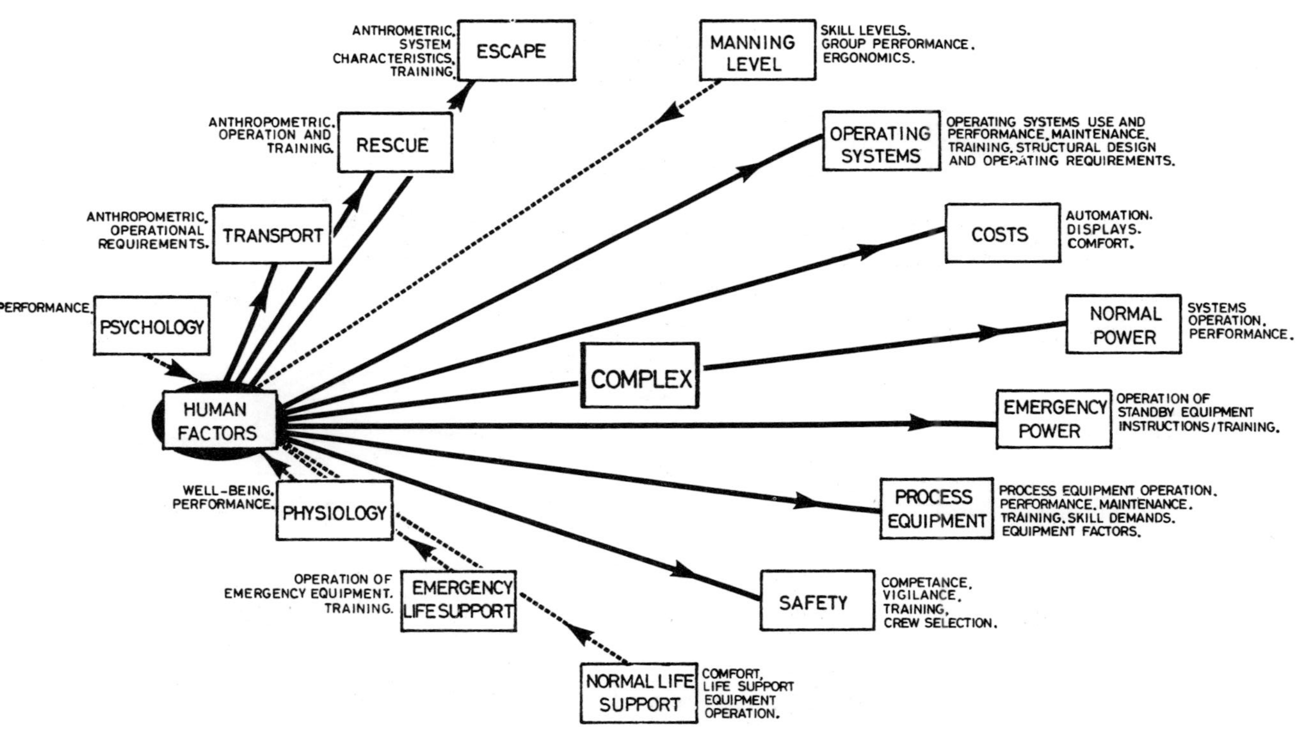

Figure 16.5 Human factors.

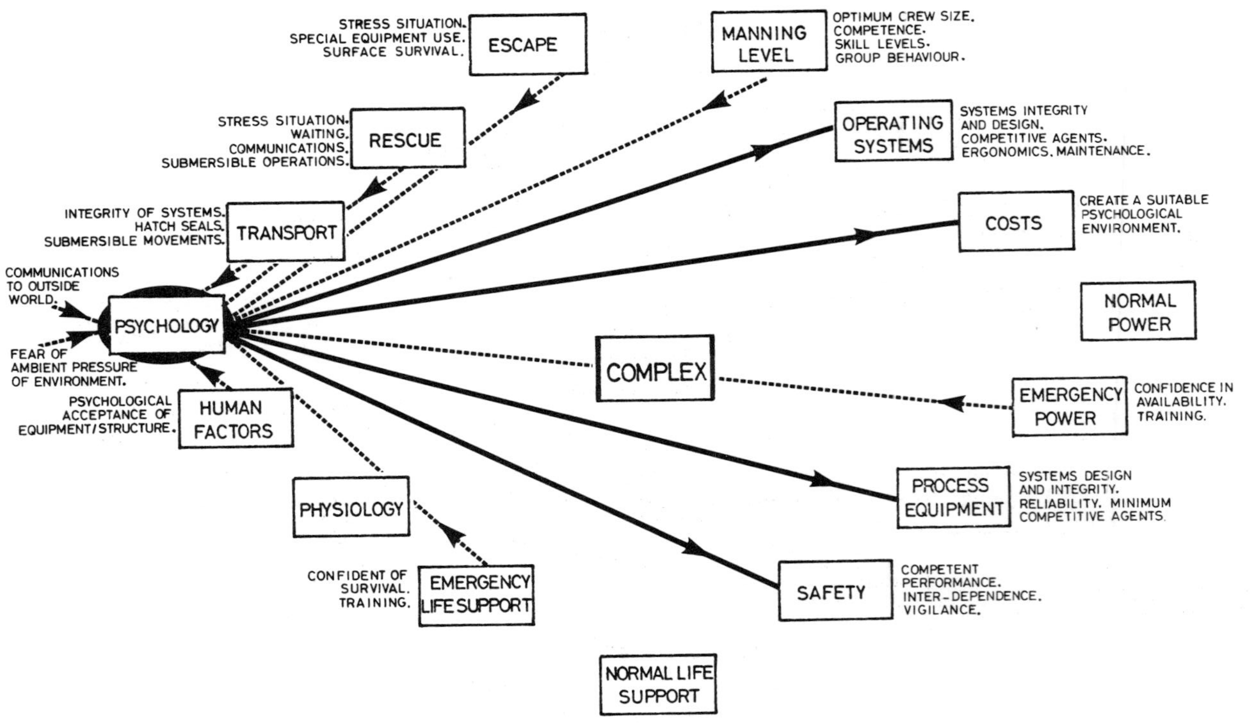

Figure 16.6 Psychology.

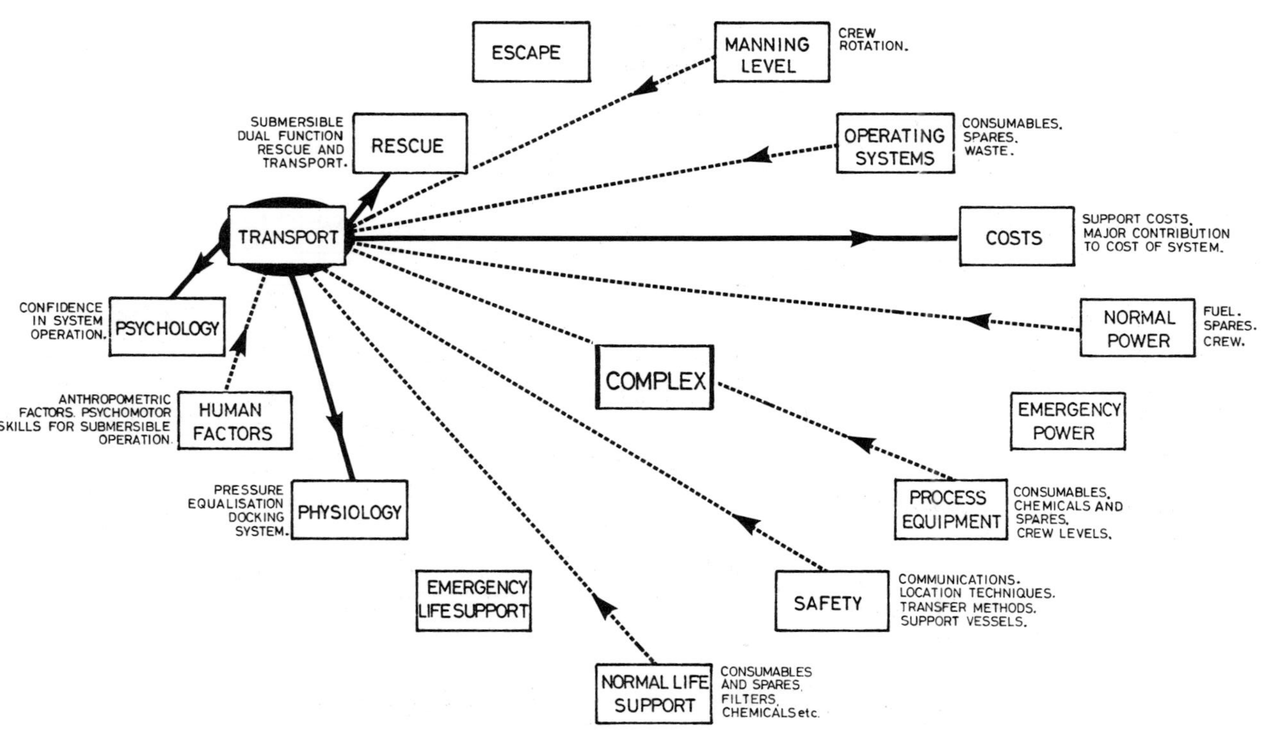

Figure 16.7 Transport.

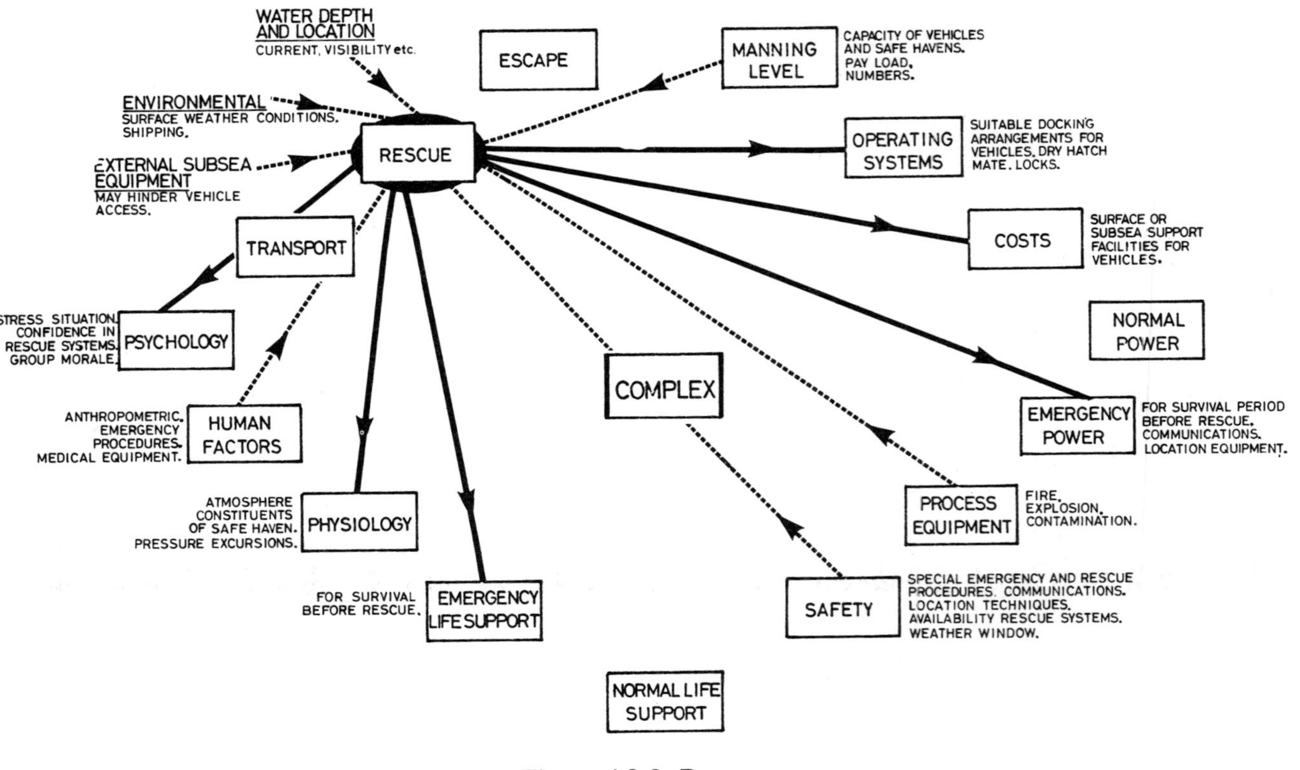

Figure 16.8 Rescue.

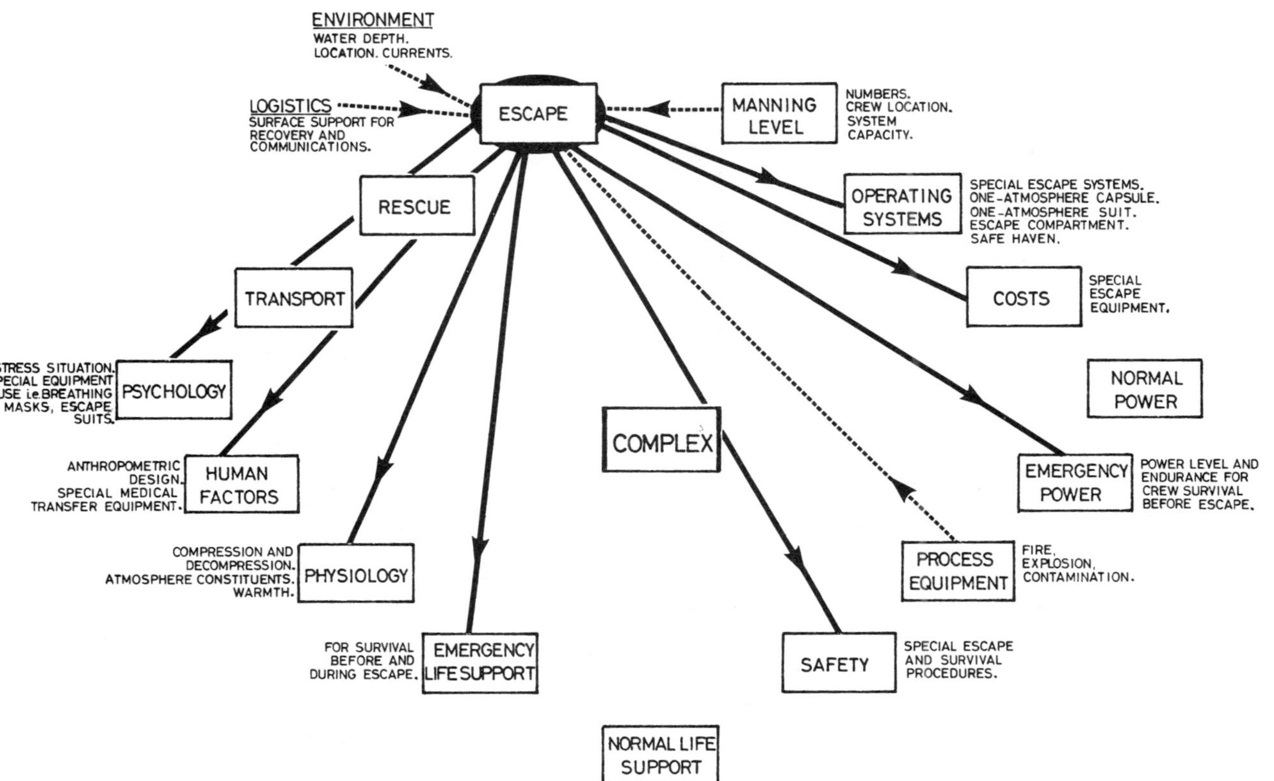

Figure 16.9 Escape.

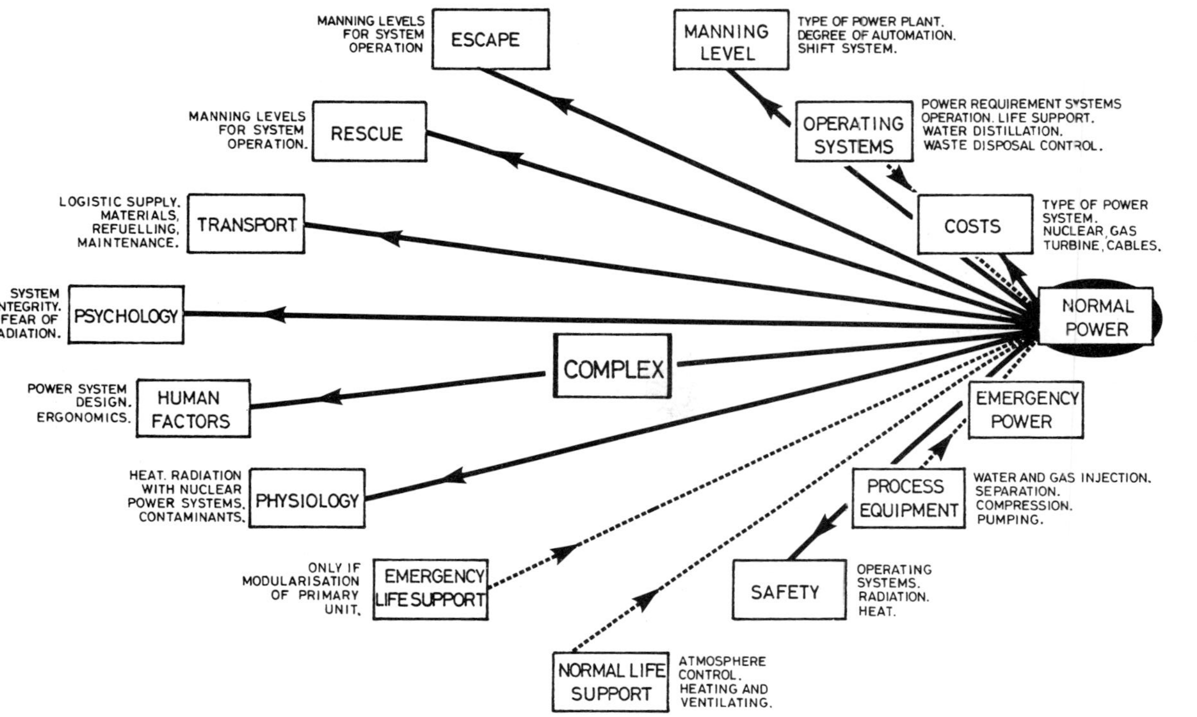

Figure 16.10 Normal power.

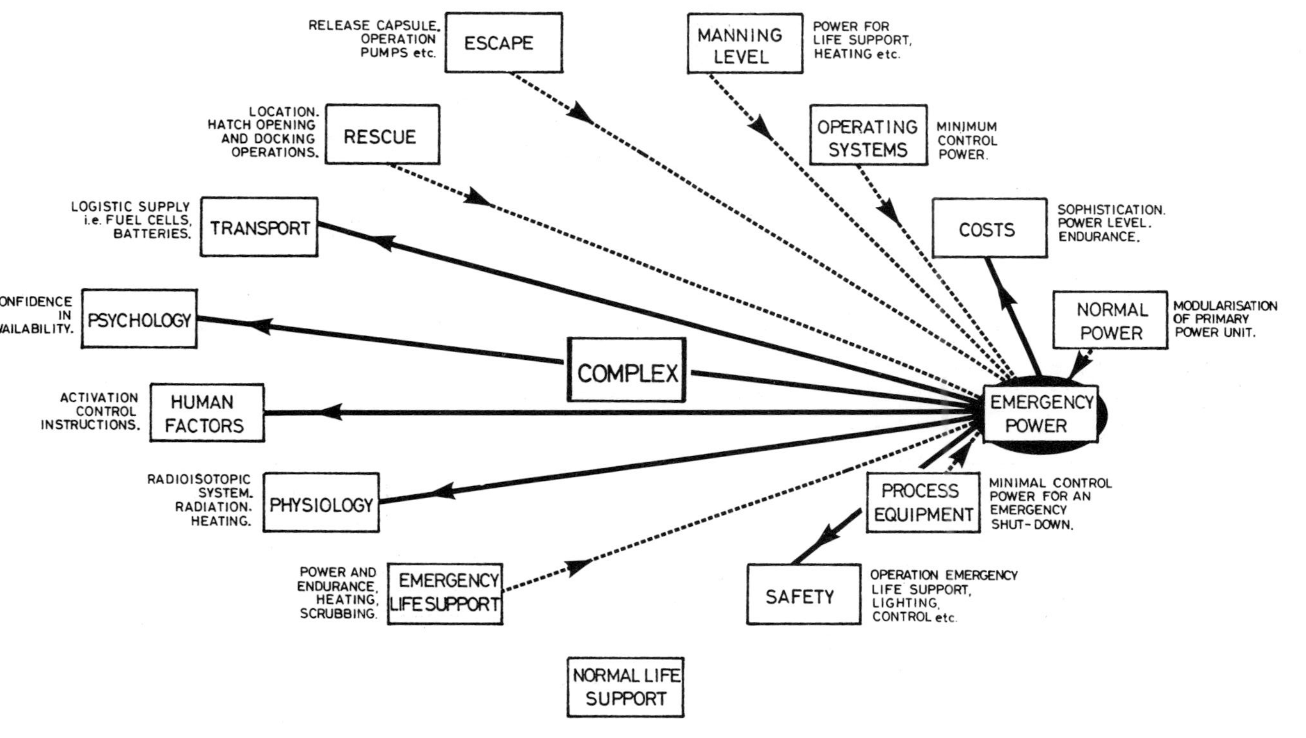

Figure 16.11 Emergency power.

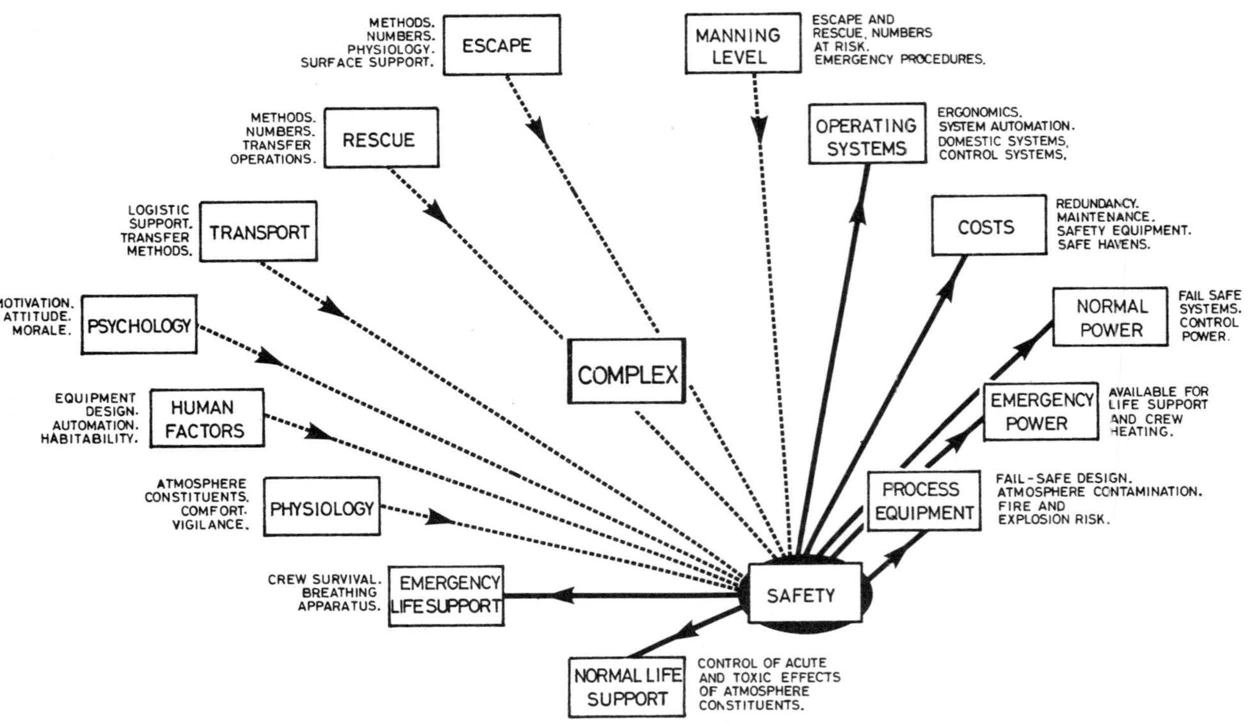

Figure 16.12 Safety.

17

RELATED TECHNOLOGY

17.1 PRODUCTION OPERATIONS REQUIRING A SURFACE INTERFACE

The exploitation of oil and gas deposits from deepwater areas offshore will probably involve production so remote from a landfall that the provision of an underwater pipeline for product export will be restrictively expensive for individual field development. In this case an interface with the surface will be required for the export of oil or gas production. It may also be necessary to flare any associated gas resulting from oil production operations, at the surface. The incorporation of such facilities in a production system will detract from the inherent advantages of the total subsea production concept. Alternative export methods may need to be considered or developed.

17.1.1 Export of Oil Products

Present systems for offshore tanker loading and product export utilize various arrangements of articulated tower or anchored buoy systems, i.e. S.P.M., S.P.A.R., etc. and it is possible that developments in these types of system may find application in deepwaters. The problems involved in such use with risers, mooring systems, etc. will, however, be similar to those encountered previously in the treatment of floating production systems. Also the economics of deepwater exploitation will probably only allow the development of fields of large reservoir volumes exhibiting high daily flow rates, resulting in a need for larger buffer storage volumes when tanker loading operations are restricted.

If an articulated tower or buoy system is incorporated into the production system to provide a supply of air or electrical power via an umbilical to the complex, it is probable that a tanker loading facility will be included in the design, along with a buffer storage system. Submarine transport of the oil product could also be considered and the size of the pressure hull could be minimized by using a pressure compensated design for the bulk carrying element of the structure. The pressure hull would only need to be large enough to accommodate the propulsion systems, control systems and crew. Such a submarine tanker system would need to operate in association with a subsea buffer storage unit. The concept would also need to be integrated with logistic supply operations and possibly power generation systems.

In the idealized situation where it is possible to correlate the development of a number of fields in an offshore area, the products from each field could be commingled and exported via a central surface loading facility or possibly by pipeline to shore. Considerable expense and electrical power for pumping will be involved in pipeline export over the long distances envisaged, however, if the costs are spread over a number of fields the concept may become practical.

17.1.2 Gas Export and Flare

The production from a gas field involves similar export problems as oil production. At present gas is normally transported from the offshore fields by pipeline, although methods are being investigated for the offshore liquefaction of gas products using a semi-submersible based liquefaction plant and L.N.G. tankers for export. Associated gas produced as a result of oil production is normally re-injected into the well or flared on the surface. Energy conservation policies may restrict flaring in the future.

Associated gas in deepwater fields will probably be flared or reinjected into the well for concentrated removal at a later date using gas production methods. Gas production fields will probably need to be developed in an integrated way with the use of a centralized surface based liquefaction plant and L.N.G. tanker export or a gas pipeline to the shore. Developments in multiphase pumping techniques (simultaneously moving oil and gas in a pipeline) should be watched as they may allow oil and gas to be pumped together in the future. This would result in reduced pipeline costs and a simplification in requirements for the subsea production complex, by eliminating the need for subsea separation and treatment.

17.2 OTHER TECHNOLOGICAL CONSIDERATIONS

Areas that are relevant to the design of the manned underwater production complex, which have not been considered previously are briefly identified as a guide to other work in the field and to indicate the limitations of this study.

17.2.1 Methods of Construction of the Pressure Hulls

The Deep Sea Production System study has considered three different material configurations for the cylindrical endcapped pressure hulls; reinforced concrete, concrete/steel sandwich and ring-stiffened steel shells. Present work suggests that at a given water depth the solution appears to lie in a method which combines the best properties of concrete and steel. For depths up to 500 m this is achieved using concrete with steel bar reinforcement. Beyong 500 m in order to maintain buoyancy composite constructions, of which the steel/concrete sandwich is a particular case, appears the most promising way of achieving a buoyant hull design for these water depths (Fig. 17.2.1). Present research work on composite structures is being undertaken in association with Sir Robert McAlpine Limited by Dr Montague and his colleagues at the Simon Engineering Laboratories, University of Manchester.

17.2.2 Oil Processing Equipment

The oil processing equipment of the D.S.P.S. complex is based on standard oil processing equipment and will be mounted on a skid for insertion into the pressure modules. Detailed design is being

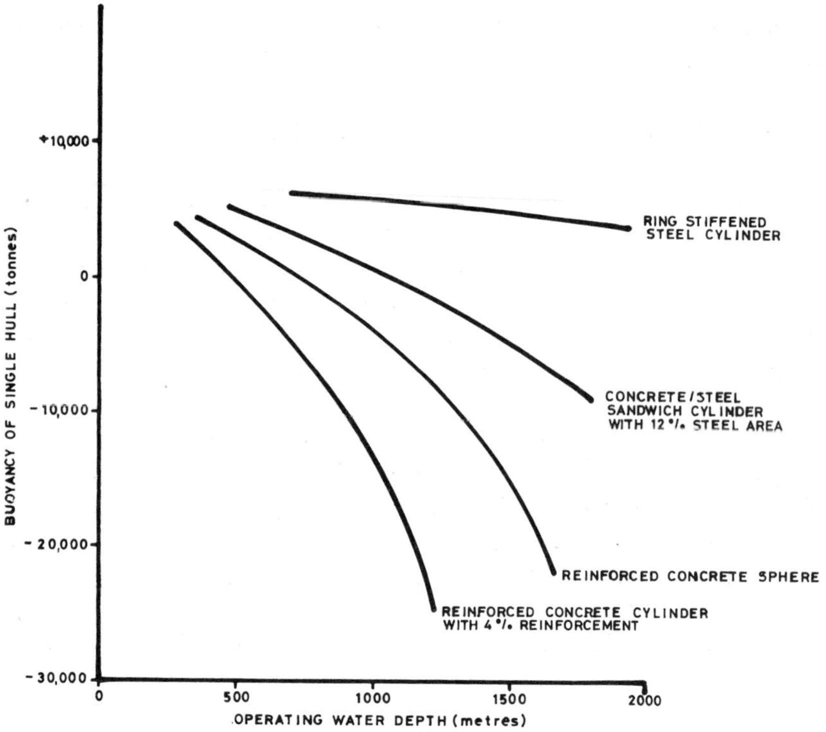

+10,000
0
-10,000
-20,000
-30,000
BUOYANCY OF SINGLE HULL (tonnes)
0
500
1000
1500
2000
OPERATING WATER DEPTH (metres)
RING STIFFENED STEEL CYLINDER
CONCRETE/STEEL SANDWICH CYLINDER WITH 12% STEEL AREA
REINFORCED CONCRETE SPHERE
REINFORCED CONCRETE CYLINDER WITH 4% REINFORCEMENT

BUOYANCY / WATER DEPTH
(CYLINDER = 12 m. dia. x 65 m. long)

4·0
3·0
2·0
1·0
0
WALL THICKNESS (metres)
0
500
1000
1500
2000
OPERATING WATER DEPTH (metres)
REINFORCED CONCRETE CYLINDER WITH 4% REINFORCEMENT
REINFORCED CONCRETE SPHERE
CONCRETE/STEEL SANDWICH CYLINDER WITH 12% STEEL AREA
RING STIFFENED STEEL CYLINDER

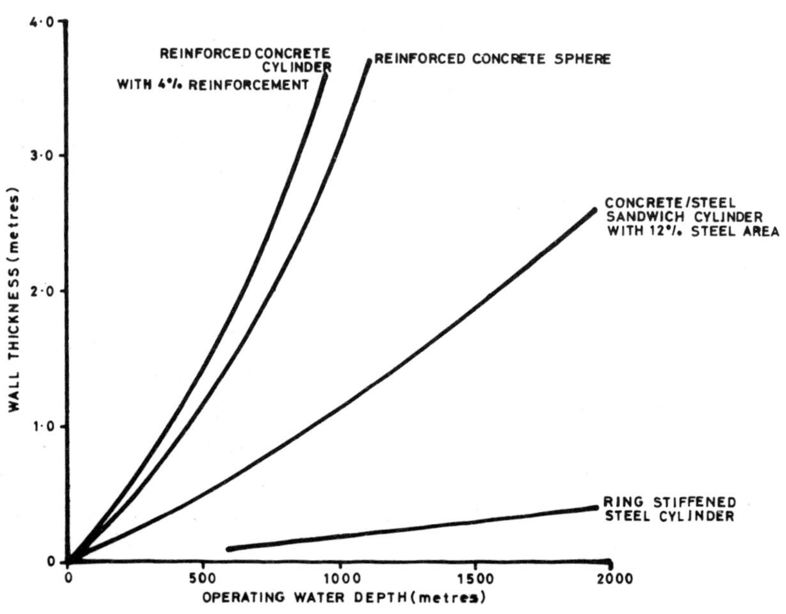

WALL THICKNESS / WATER DEPTH
(CYLINDER = 12 m. dia. x 65 m. long)

undertaken by Humphreys & Glasgow, a member of the D.S.P.S. consortium.

17.2.3 Fabrication, Transport and Installation

The pressure hulls and buoyancy cells will be built in a purpose made dry dock and the draught of the finished structure will be such that completion and test can be carried out in the dry dock. The plant packages will be prefabricated and skidded into the partially completed cylindrical pressure modules, final hook-up and testing will be carried out during construction of the dome closures of the cylinders.

The dry dock will be flooded, the entrance removed and the complex towed to the deepwater location, where it will be water ballasted to the required buoyancy and lowered or pulled to the seabed using surface vessels and winches. Construction would be undertaken by Sir Robert McAlpine (D.S.P.S.) and studies on practical immersion techniques have been carried out in association with Noble Denton and Associates and Netherlands Offshore Company.

17.2.4 Power Supplies and Umbilicals

Rolls Royce Limited have undertaken feasibility studies into the use of gas turbine generators and subsea nuclear reactors for power generation for the complex. British Insulated Callender's Cables Limited have carried out studies on the provision of umbilical connections from the surface to the complex and transmission cables from the shore to the complex. Both companies are members of the D.S.P.S. consortium.

17.2.5 Ocean Technology

The geological parameters and oceanographic environmental constraints will impose hazards on the manned underwater structure. The water column will be a major obstacle during phases of complex emplacement and logistics support operations. Communications will be complicated by the water column. The water/seafloor interface characteristics will influence structure foundations and design.

Oceanographic features such as temperature, salinity and oxygen content will need to be determined. Currents that flow along the ocean floor will have considerable importance in the design of the structure, and shallow currents will affect emplacement techniques. Catastrophic occurrences such as seismic and volcanic activity must also be considered in the evaluation of a location and the design of the structure. A bottom-sitting underwater structure will receive at least partial support from the bottom sediments; soil mechanics investigations will be needed to establish whether the load can be supported. The degree of flatness of the ocean bottom and angle of inclination of the seabed will influence levelling techniques required for structure location.

Biological and thermal pollution of the ambient environment surrounding the structure will be of concern. Waste disposal will need to be controlled and remote from the structure, thermal pollution will encourage marine growth and complicate submersible operations by creating underwater thermals and possibly temperature gradients. Potential hazards to submersible operations due to subsea flowlines and ancillary equipment external to the complex, will need to be removed at the design stage. Subsea equipment that will require servicing or maintenance *in situ* must be designed to interface with submersible or diving equipment used for the task.

The integrity of communication and location systems between the complex, submersibles and surface support facilities in association with proper operating procedures and accountable command structures will be key elements in the safe operation of the complex and peripheral systems.

18

FUTURE SUBSEA OIL PRODUCTION CONFIGURATIONS

A tentative attempt is made to identify some alternative subsea production configurations, based on possible developments that have been suggested in the previous analysis. The full technical implications of each configuration cannot be reviewed in a work of this nature; however, various scenarios are generated to indicate the direction in which future deepwater offshore operating systems may develop.

A simplified system drawing is used to present each configuration, major subsystems are identified and alternative subsystem options indicated. A brief discussion of each system is given to clarify the selection of the various parameters.

The system configurations are developed progressively from hybrid subsea systems, relying on a relatively high level of surface support towards the more truly subsea system. The final configuration considers the idealized situation, where several deepwater offshore fields are developed as an integrated system. The Exxon Subsea Production System could also have been considered, as this automatic subsea production system may provide a viable alternative to manned underwater operations in deep waters.

In all configurations, it has been assumed that beyond the effective diver working depth (300 m) subsea completions will be dry and at one atmosphere. Encapsulated equipment maintenance and repair will be undertaken by manned intervention from a submersible or capsule. Ambient pressure systems maintenance will be undertaken by submersibles specially interfaced for the task, or JIM type systems. The JIM type systems will be locked out of the subsea complex for local tasks or deployed from large submersibles or submarines for work remote from the complex.

The application of a specific system configuration will obviously be determined by economic and practical limitations. Environmental conditions, i.e. water depth, location and distance from a landfall, will be key considerations. Economic viability will depend on the field characteristics and market forces in relation to development costs.

18.1 SURFACE POWER GENERATION, AIR SUPPLY AND PRODUCT EXPORT

Figure 18.1 identifies the major components of the configuration. An articulated tower provides a surface control centre and carries power generation equipment and air pumps. The electrical power and a breathable air supply are fed to the complex via umbilicals. A surface support vessel acts as stand-by vessel and provides launch and recovery facilities for submersible operations.

The major danger to the complex is the possibility of the loss of the electrical or, more importantly, the air umbilical from the surface to the complex. Tanker loading and submersible operations will be weather dependent and gas turbine generator exhausts may produce dangerous thermal flows affecting helicopter landings on the platform.

In the event of the loss of umbilical connection, it would appear advisable to provide a relatively sophisticated emergency back-up system. A radioisotopic generator will provide emergency power that can be continuously available and of sufficient endurance to meet requirements. If an air umbilical was to become unserviceable a comprehensive emergency life support system would be required to support the large crew numbers during the fault condition. The advanced state of development of life support systems for submarine and space systems may justify the installation of a full life support system for the complex initially, in preference to risking the complications of an umbilical air connection. The provision of an *in situ* life support system would have inherent safety advantages.

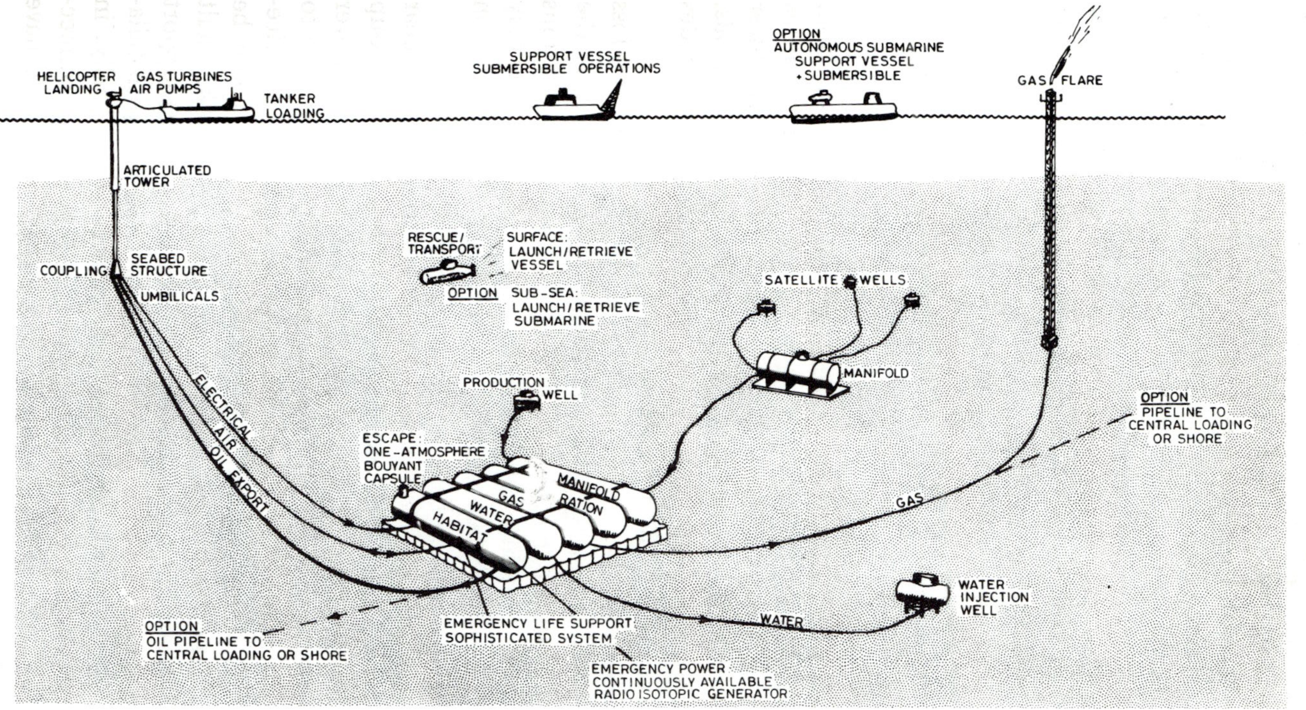

Figure 18.1 Surface power generation, air supply and product export.

The problems of product export can be alleviated by the use of buffer storage, either in the tower or within a subsea unit, although transportation via a shared pipeline to shore may be possible. The limitations imposed on submersible operations due to their weather dependence, could be a major restriction. Consideration could be given to the design of an autonomous submarine that could act as a stand-by vessel and submersible base. In bad weather the submarine could dive to avoid the surface conditions, and perform subsea transfer operations between the complex and submarine using a submersible riding piggy-back on the submarine hull. The submarine support vessel could return to port when replaced by a similar vessel, the submersible being transferred between each vessel subsea.

18.2 SUBSEA PRODUCTION COMPLEX WITH ARTICULATED COLUMN

Figure 18.2 attempts to combine the features of the subsea production complex with the features of the articulated buoyant column design Arcolprod (Taylor Woodrow). Present buoyant column design studies are centred on a limited depth range, i.e. 200 - 400 m, although it may be possible to extend this range. It has been assumed that by simplifying the design, i.e. removing drilling and production operations, a design concept may be possible that is compatible with the subsea complex.

The cellular concrete base of the tower, which is anchored to the seabed is assumed to incorporate a one-atmosphere access way to the habitat module of the complex. The design features of the tower (see Section 3.3.1) include a one-atmosphere access shaft from the surface plinth down to the cellular base, this could be used for personnel and material transport and export of product for tanker loading on the surface. Power generation and forced ventilation could be provided from the surface plinth and fed via the access-shaft and access-way to the complex. The tower could also provide buffer storage for tanker loading operations.

A tower of such design would have improved safety characteristics over the simple articulated tower utilizing umbilical connections, such integrity in the system design may only call for minimal emergency facilities. Materials and personnel transport to the

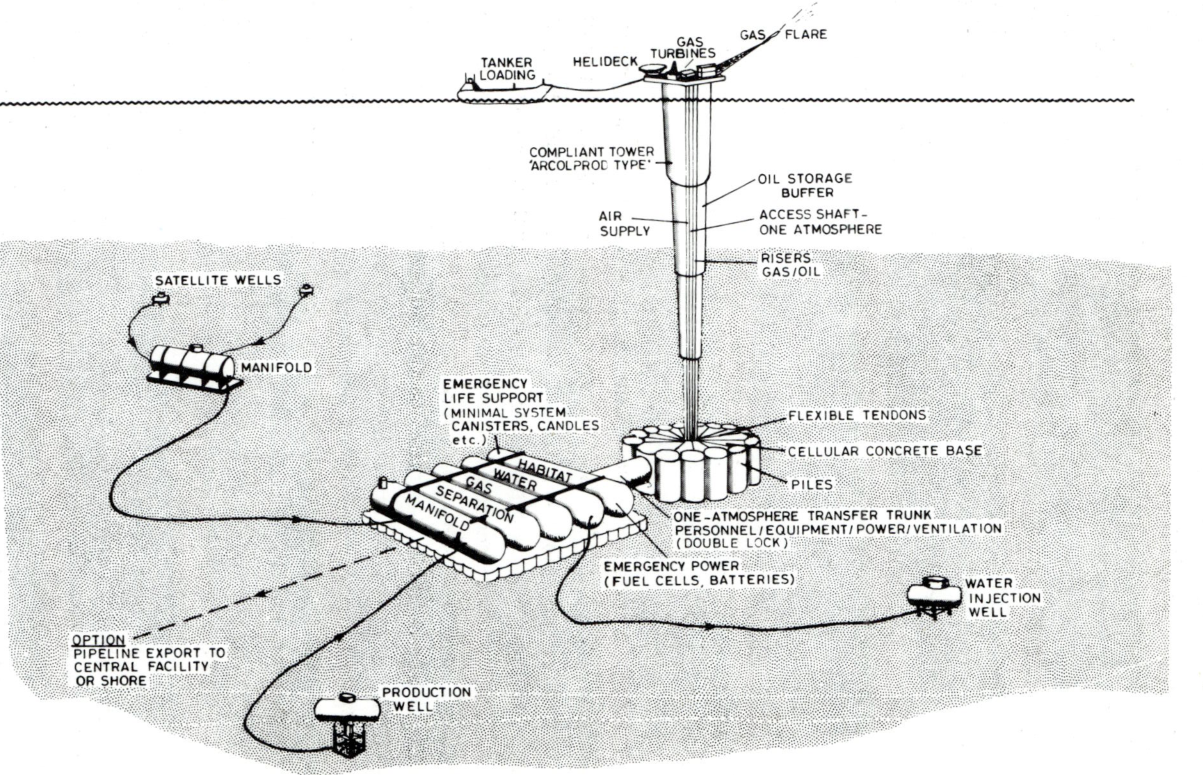

Figure 18.2 Subsea complex with articulated column.

complex and rescue would be independent of surface or submarine operations. Power generation and life support could be provided by conventional surface systems.

The combination of these two systems appears to provide an ideal solution to many of the inherent problems of the subsea system; however, the feasibility of such a concept will require detailed investigation. The possibility of designing a simplified tower concept for the water depths envisaged and an access-way between the subsea complex and the cellular base are open to question. The combined costs of the two structures may be uneconomic, and the full development of either concept may be more cost effective.

18.3 TOTAL SUBSEA PRODUCTION CONCEPT

A conceptualized scheme for a total subsea production system is illustrated in Fig. 18.3. The only surface piercing component of the system is a flare tower for associated gas flaring. The tower would also act as a surface marker for the otherwise totally subsea system and may incorporate an umbilical air connection to the complex. The umbilical would only be used for intermittent venting of the enclosed atmosphere in the event of a problem with long term build-up of trace contaminants. The complex would contain a full life-support system based on submarine and space technology with an emergency back-up system. Logistic support would be provided by an autonomous submarine acting as a base for submersible operations. Escape facilities would consist of one atmosphere buoyant ascent capsules.

The major restrictions to the development of such a concept are the requirements to export oil and gas products from the field and the provision of adequate power levels subsea for operating systems. Nuclear reactors are the only subsea power source capable of generating the required power levels (MW) *in situ*. The reactors in the concept are assumed to be Rolls Royce pressurized water reactors or Rockwell International Sodium/Potassium liquid metal cooled reactors. The reactors may be manned or unmanned depending on the operating philosophy, but redundancy will probably be provided with the multi-module configuration indicated. The inherent redundancy in the primary power supply

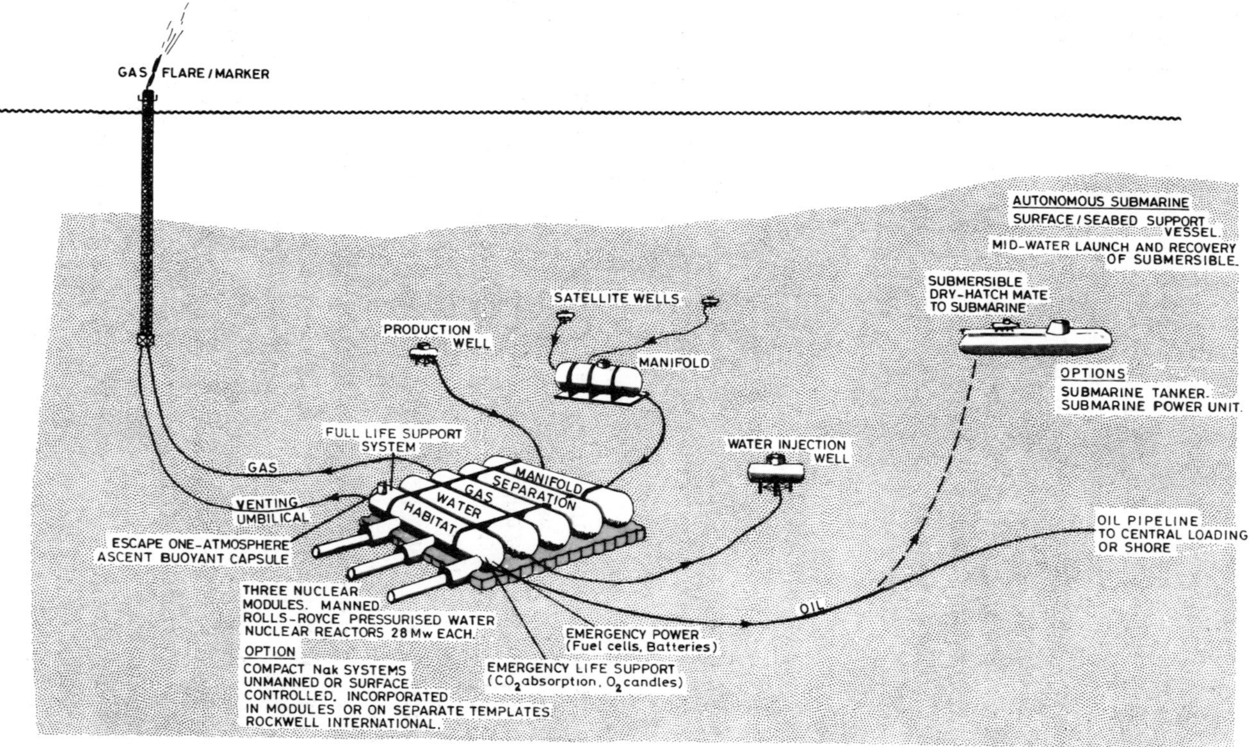

Figure 18.3 Total subsea production concept.

only requiring the provision of batteries or fuel cells for the emergency power system.

The ideal method for transportation of the product from a totally subsea production system would be a submarine pipeline to a landfall. The distance from shore, water depth and seabed gradients involved may make this method impractical for a single field development. The integrated development of a number of fields in a remote offshore area may justify the expense of a communal pipeline to shore. Future developments in multi-phase pumping techniques may allow simultaneous export of oil and gas in the same line, avoiding duplication of pipelines and reducing the complexity of subsea production systems by eliminating the need for separation and treatment of the product. Submarine tankers could also be considered for oil transport, but would need to operate in association with a large subsea buffer storage system. The functions of the tanker would also need to be correlated with logistic support operations and power provision for the complex.

If multi-phase pumping techniques do not materialize, excessive costs restrict the use of pipelines or submarine tankers are impractical, consideration could be given to the use of a central oil loading or gas liquefaction plant to service the production of several fields. The aim being to minimize the number of surface interfaces in the production area.

18.4 MULTI-FIELD DEVELOPMENT CONCEPT

The development of the concept is based on the experience that production fields are normally located in a relatively restricted geographical area. The separation distances between adjacent fields in a deepwater offshore production area may be small, compared to the distance to the nearest landfall. The concept considers the systematic development of a number of fields in a geographical area as an integrated unit, in a similar manner to the development of a single field using satellite wells. The difficulties inherent in such an approach are appreciated, i.e. correlation of field development, company politics etc. but this approach may be the only viable way of developing deepwater deposits economically.

The block diagram in Fig. 18.4 shows a proposed layout of a multi-field configuration. The nucleus of the total system is a

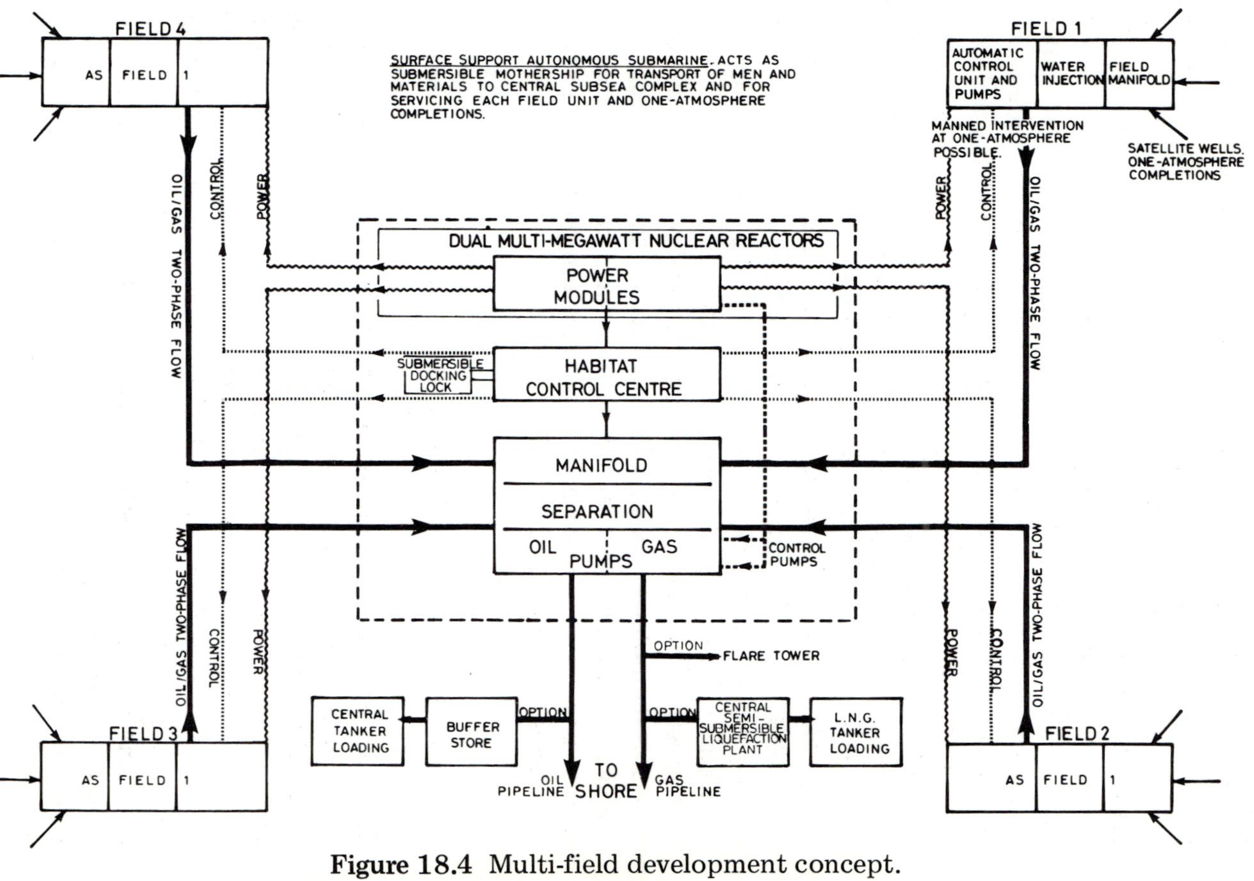

Figure 18.4 Multi-field development concept.

collection of one atmosphere pressure modules incorporating power generation, central processing, a habitat and a control centre. It has been assumed that future developments in multi-phase pumping techniques will allow the transport of oil/gas from each 'satellite field' production centre to the central processing unit over the relatively short distances, but is unlikely for the distances and gradients to a landfall.

Each 'satellite field' production centre consists of one atmosphere modules containing: a manifold to commingle production from the satellite wells, water injection, pumps and a control unit. Gas treatment and separation are not required if multi-phase pumping to the central complex is assumed. The field centres would normally be unmanned and controlled automatically from the central control unit. One atmosphere manned intervention to each field centre and dry satellite wells would be possible by submersible for maintenance and repair operations. The field centres would obtain power for control and pumping by undersea cables from the central power module.

The central processing module would commingle the production of the 'satellite fields' and perform gas separation and treatment and export of the products. The large volume of products would probably justify the expense of a pipeline for gas and oil export, otherwise a central loading facility with a large buffer store could be used for oil export and a surface liquefaction plant for gas export.

The central habitat and control centre would provide manning for 'satellite field' control, reactor operation and central processing systems. The power generation system would probably consist of dual multi-megawatt nuclear reactors providing redundancy and cost effective operation at the high power levels envisaged. Variations in field power requirements over the life of the fields will allow optimization of reactor design. Sufficient power will be available to allow the products to be pumped to a landfall if pipelines are feasible.

An autonomous submarine could again be used as a stand-by vessel and as a mothership for submersible operations. The submersible would be launched from the submarine to perform 'dry-hatch' docking for intervention to field production centres and dry satellite wellheads. The scale of the central unit may justify the inclusion of an underwater lock to provide docking of the submersible in a one atmosphere environment subsea.

The problems of developing an offshore production area in such a manner may be unresolveable, and in many senses impractical. The integrated field development concept is advantageous, because the costs of major subsystems can be spread over a number of field developments. This concept may be the only means of providing commercially the technologically sophisticated subsystems, i.e. autonomous submarines, underwater reactors, deepwater pipelines etc., that could be required for the exploitation of deepwater resources.

18.5 SUMMARY

The feasibility of exploiting deepwater deposits of oil and gas will depend on the discovery of reservoirs of sufficient capacity and wells of high daily flow rates to justify the elevated cost of development of such fields. New production technology will have to be developed for operations in the deeper and more hostile environment. The high financial risks involved in the introduction of new technology will predetermine the development of innovative systems. Only when existing proven systems cannot match contemporary requirements will new systems be incorporated. Total subsea production systems will therefore probably be developed in a hybrid fashion, initially highly dependent on surface systems and evolving into the total subsea system when required by the industry.

At present the major restrictions to the provision of a totally subsea production system in deepwater are the need to interface to the surface for product export and power generation. Logistic support operations for the complex will also be limited by environment conditions on the surface.

Pipeline export is the ideal solution for product transportation; however, the distance and gradients to a landfall from deepwater deposits may make the provision of a pipeline restrictively expensive for a single field development. A communal pipeline for commingled export of production from several fields may be justifiable, especially if multi-phase pumping techniques evolve.

Submarine tankers could be considered although they may not be able to cope with the high daily flow rates. If tanker loading proves necessary, to minimize the number of surface interfaces, a central loading facility with a suitable buffer subsea storage could be used to export the product of several fields.

The provision of megawatts of electrical power *in situ* in the underwater environment is technically feasible using nuclear reactor powerplants. The operational problems and costs for a single field development would probably be unacceptable, unless pipeline export of product is possible and high power is required for pumping. A central nuclear reactor complex could probably be justified for a multi-field development concept.

The concept of an autonomous submarine acting as a surface or seabed stand-by vessel and mothership for submersible operations has many distinct advantages over present day systems. The cost of such a vessel will be high but the operational gains would be significant, especially if other subsea operational requirements could be integrated into the design, i.e. product export, power system.

There are no technical feasibility problems with life support systems or power supplies. Full life support systems are considered appropriate even where air umbilicals have been suggested. Fuel cells or batteries are suitable for emergency power supplies, except where an electrical and air umbilical are used when a radioisotopic generator is recommended. The preferred method of rescue from the complex is by submersible transport and escape by the buoyant one atmosphere capsule, where these facilities are required.

18.5.1 Future R & D Requirements for the Exploitation of Deepwater Resources

1. Investigate the use of improved articulated columns and buoy loading systems for deepwater exposed locations.
2. Investigate the problems in achieving a reliable riser system for deepwater production and loading operations. Related problems of umbilicals, diverless subsea connections, subsea control and riser handling should also be considered.
3. Assess the present technology of multi-phase pumping techniques for the simultaneous transport of oil and gas and

possible future developments in relation to pipeline distances and gradients.

4. Pipeline studies on the feasibility of exporting products from deepwater locations to a land-fall by considering laying techniques, protection and pumping power requirements.
5. Investigate the state-of-the-art of nuclear powerplant technology for subsea application. Nuclear reactors for high power requirements (MWe) and radioisotopic generators for lower powers (kWe).
6. Investigate other potential subsea power sources, batteries, fuel cells, thermo-electric etc., and establish the state-of-the-art.
7. Establish the technology of undersea electrical transmission systems and underwater electrical connectors.
8. A feasibility study should be undertaken into the possibility of developing an autonomous submarine for use as a stand-by vessel and submersible mothership. The concept should also consider the options of a submarine tanker and a submarine mobile power unit.
9. Investigate the use of JIM type systems for subsea maintenance tasks associated with the complex in deepwater, i.e. external work tasks to the structure, inspection of structure, pipeline and completion maintenance. The logistics of docking such systems out of the structure, from a large submersible or submarine should also be considered.
10. Investigate the design of external equipment and submersible systems to provide compatible interfaces between the systems for maintenance and repair tasks.

Costs

It has not been possible in a work of this breadth to investigate the costs of system configurations and their subsystems. Section 5.1.4 from Ref. 55 does provide some general conclusions on costs from the studies of manned underwater stations. Details of capital and operating costs for the subsea complex considered in this work (D.S.P.S.) are given in Ref. 48.

SECTION D

Conclusions and Recommendations

19

CONCLUSIONS

1. Significant prospects of finding commercially exploitable deposits of oil and gas offshore exist on the continental margins of the Atlantic, Indian and Arctic Oceans. The deposits will be difficult and expensive to find and extract and may involve operations in water depths approaching 2000 m by the end of the century.
2. The exploitation of resources in such deepwaters will require new and expensive technology, which will probably advance by staged development rather than by a technological breakthrough, because of the high financial risks involved. Special deepwater drilling technology is at present evolving and should not be a restriction to the development of deepwater deposits. Riser systems are at present the weak link between sophisticated subsea and surface systems. The basic concept of subsea completions, manifolds, templates and T.F.L. tools is well established.
3. Present bottom fixed production systems will probably be limited to 200 m water depth in harsh environments and 300 m in more sympathetic environments. The guyed tower will have application in water depths in 300 - 600 m range. The tension leg concept may extend operations to 750 -1000 m, although beyond 500 m its application will be dependent on advances in the technology of riser design,

mooring systems and diverless subsea systems. At depths greater than 1000 m only the Deep Sea Production System (D.S.P.S.) and possibly the Submerged Production System (Exxon) would appear to have the necessary potential for this application. The Exxon S.P.S. system has undergone engineering field trials while other systems are conceptual.

4. In the specification of suitable production systems for various water depth ranges a contentious area exists between 500 – 1000 m. The use of total subsea production systems or surface piercing structures at these depths, is an area of uncertainty. The limitations of tension leg concepts and automatic subsea production systems, indicate that detailed studies of the manned subsea concept are justified for water depth applications from 500 to 2000 m. The concept may also be applicable to shallower water depths once established and may also have other applications in the oceanographic field.
5. Past manned underwater structure design studies indicate that there appear to be no crucial technical feasibility problems in the implementation of the concept. Submarine, space and diving technology provide valuable technical information for the design. The use of one atmosphere habitats for subsea completions and manifolds in the oil industry is established and can provide useful data.
6. Generally, the technology exists to provide environmental control and life support for the concept. There will, however, be many environmental problems unique to the application as an oil/gas production complex. The permanent emplacement on the seabed of the complex may involve contamination problems beyond the present state of knowledge.
7. The provision of a suitable physiological environment for crew maintenance can be achieved using present technology. The crew will be operating on a two-week rotation schedule, this may ease limitations on maximum permissible concentrations of contaminants during their enclosure.
8. Habitability and systems design will impact crew performance and affect the psychosocial climate within the structure. Human factors will have to be evaluated in the complex design.
9. It is unlikely that any overt psychological disturbances will occur amongst the crew. The use of suitable crew selection and

training criteria in association with the satisfaction of human requirements will minimize stresses imposed on the crew.

10. The logistic support of the manned underwater structure involves several problem areas and will be crucial to the successful day-to-day operation of the complex. Present surface support systems and dry transfer methods may impose limitations on the use of large submersibles for support operations. A purpose-designed submersible for the tasks will probably be required, operating from an autonomous submarine mothership to alleviate problems encountered at the surface interface. One atmosphere diving suits and submersibles will probably be used for external subsea maintenance tasks.
11. Submersible transport of personnel from the structure in combination with an acceptable support system is preferred as the primary means of rescue. The incorporation of one atmosphere submarine escape capsules within the structure for escape is recommended as a secondary system, should submersible operations be terminated. Consideration could also be given to the use of one atmosphere diving suits for escape by individual buoyant ascent in isolated areas of the complex remote from central escape facilities.
12. The development of deepwater fields is likely to be at a distance from the shore, the use of transmission cables for power supply may be impractical. Surface generation and supply by umbilical may be unreliable. A subsea nuclear reactor is the only feasible means of generating megawatts of power underwater at the field location. For emergency power supplies (kW) fuel cells or batteries are recommended with non-umbilical primary power supplies and radioisotopic generators for systems employing umbilical feed primary power supplies.
13. Safety considerations will affect all operational systems, crew exposure to hazards within the station will be continuous. The crew will also be exposed to additional hazards during crew transfer operations or escape and rescue procedures. Detailed safety analysis during design and manufacture should reduce the level of risk to the crew to an acceptable level for industrial safety.
14. Developments in related areas of technology will need to be assimilated in the design studies. The major limitations to the

provision of a totally subsea production system in deepwater are the need at present to interface with the surface for product export, power generation and logistic support operations. Technological developments in these areas will need to be encouraged. Improvements in deepwater loading systems, pipeline transportation, pipe-laying techniques, multi-phase pumping and other ocean technology systems are required.

15. The complex interactions that occur between subsystems in such a design concept will need to be evaluated in detail, once a full system configuration has been finalized. It has not been possible to perform this analysis in this study, although some considerations have been indicated.

16. It is considered that beyond the effective diver working depth (300 m) subsea completions and manifolds will be dry and at one atmosphere. Encapsulated equipment repair and maintenance will be undertaken by manned intervention using a submersible or capsule. Ambient pressure systems maintenance will be undertaken by submersible or JIM type systems specially interfaced for the task requirement.

17. Total subsea production systems will probably develop in a hybrid fashion, initially highly dependent on surface systems and gradually evolving into a totally subsea system as existing proven technology fails to meet the contemporary requirements of the industry. The technologically sophisticated hardware that may eventually be required for the exploitation of deepwater resources, i.e. autonomous submarines, underwater reactors, deepwater pipelines etc. may only be justified by the integrated development of an offshore area.

20
RECOMMENDATIONS

The following areas have been identified as requiring further research and development to form a suitable technological base for the use of manned one atmosphere underwater structures for the exploitation of deepwater resources:

1. Investigate the possibilities of the use of improved designs of articulated columns and buoy loading systems for deepwater (500 m+) exposed locations. Riser and mooring design problems should be evaluated and related requirements for umbilicals, diverless subsea connections, subsea control and riser handling considered.
2. Investigate the feasibility of using subsea pipelines for product export from deepwater locations to a landfall. Distance, gradients, laying techniques, protection and pumping power requirements should be considered. The present technology and possible future developments in multi-phase pumping techniques should be assessed and assimilated into the study.
3. Investigate in detail potential subsea power sources and establish the state-of-the-art. Nuclear reactor powerplants should be investigated for high power requirements (MW). Radioisotopic powerplants, fuel cells, thermo-electric systems and batteries should be investigated for lower power levels (kW). Developments in nuclear plants for space and maritime applications during the last decade should be especially relevant.

The technology of undersea electrical transmission and underwater connectors should also be investigated.

4. Investigate the use of one man, one atmosphere submersibles (JIM type systems) for ambient subsea maintenance tasks associated with the operation of the complex. The tasks will probably involve external inspection of the complex structure and cable, pipeline and completion maintenance and inspection. Attention should be given to a design, producing a compatible interface between the submersible maintenance vehicle and the equipment on which the work is to be performed. The logistics involved in deploying JIM type systems, from the structure, from a large submersible or submarine should also be considered.
5. Investigate the provision of support logistics for the undersea complex. Submersible operations involved in the transfer of men and materials to the structure, and the submersible's role as a rescue vehicle should be considered. Docking, mating, communications and locating techniques should be evaluated along with the limitations of present surface support systems. One atmosphere escape capsules may also need to be integrated with rescue systems design.
6. A feasibility study should be undertaken into the possibility of developing an autonomous submarine for use as a surface or seabed stand-by vessel and mothership for submersible operations in support of the complex. The options of the submarine acting as a tanker or mobile power unit should also be considered.
7. Investigate in detail the atmosphere requirements for the underwater structure, to identify specific problem areas and establish the appropriate technology level such that a suitable work environment may be produced. Data on acceptable atmosphere compositions should be gathered with special reference to maximum permissible concentrations of contaminants, toxicity and explosive limits. Heat loads from equipment should be considered and environmental control equipment and measuring instruments evaluated.
8. The physical implications of atmosphere composition may need to be tested experimentally in a manned test facility to determine the physiological effects of maximum permissible concentrations on the crew during two-week tours of duty.

Human factors research will be required at the initial design stage.

9. Safety analysis will be required during design, manufacture and operation of the system.
10. Once a full system concept has been finalized, a matrix analysis of the system configuration will need to be undertaken to identify all interactions between component systems, to allow optimization of the total system design.
11. Present and proposed production configurations should have their operational capabilities continuously assessed in the light of past operating experience and possible future applications. Field experience of the Exxon S.P.S. system should be continually evaluated as this system may provide a viable alternative to manned subsea production operations.
12. As an initial step towards the development of a manned underwater complex, it is considered that the construction of a one atmosphere manifold centre should be undertaken in the near future. Such a limited application of the concept would provide operating experience and a valuable information base for the design of a subsea production complex.

ACKNOWLEDGEMENTS

I would like to acknowledge the assistance of the following people during this work, by discussion of the concept and in some cases the provision of information.

Mr J Derrington Mr M Collard Mr D Price	Sir Robert McAlpine & Sons Ltd
Mr A. Dorr Mr D Kemp	Humphreys & Glasgow Ltd
Mr R Whitfield Mr J Hurst Mr G Salt Mr P Jones	Rolls Royce Ltd
Dr J Birks Mr D Thornton Mr K Searle Mr A Webb	British Petroleum Ltd
Mr L Allcock	Shell (UK) Ltd

Mr R Street Mr J Hargreaves Mr N Mansfield Mr Mayo	Department of Energy (P.E.D.)
Mr G Crouch	Offshore Supplies Office
Mr G Lisley Mr B Hindley Mr S Matthews	Health & Safety Executive
Mr D Dunkason	Marine Technology Support Unit (MaTsu)
Mr Christopher	Admiralty Experimental Diving Unit
Mr D McKellar	Y-ARD Limited
Dr M Hargreaves	University of Newcastle
Mr A Fyfe Mr A Lyons Mr A Myres (Librarian)	Institute of Offshore Engineering
Mr A Sherwood (Librarian)	Institute of Naval Medicine
Mr J Barrack	Worley Engineering
Mr W Walker	Lockheed Petroleum Services, Vancouver
Mr J Vadus	United States — N.O.A.A.
Mr T C Wang	Harbour Branch Foundation (US)

Special thanks go to Professor J Brown of the Department of

Petroleum Engineering at Heriot-Watt University and Dr D M Davies of the Occupational Health Centre of British Petroleum Limited for the constructive criticisms of this book.

Credit should be given to Isabelle Lawson for the preparation of the illustrations for the report and to Nancy Jones for typing the final manuscript, as without their patience and understanding it may not have been completed.

Acknowledgement is due to the following organizations for granting permission to use our artist's impression of their systems and for the reproduction of various tables:

COMPANY	*FIG. NO*	*TITLE*
B.P. Limited	1.1	Generalized Scheme of Hydrocarbon Generation
B.P. Limited	1.2	Schematic Cross-Section through a passive Continental Margin
B.P. Limited	1.3	Continental Margin Types
B.P. Limited	1.6	Deepwater Drilling Activity
Scott Lithgow (Offshore) Limited	3.1.1b	Fixed Steel Gravity Platform
Shell (U.K.) Limited	3.1.1c	Shell/Esso Brent B Platform Condeep Design
Norwegian Contractors	3.1.1d	Condeep T 300
Maschinenfabrik Augsburg-Nurnberg	3.1.1e	M.A.N. 400

Advanced Production Systems	3.1.2a	Floating Production System
Exxon Production Research Company	3.2.1	Guyed Tower
B.P. Limited	3.2.2	Tethered Buoyant Platform
Gulf Research and Development Company	3.2.3	Tethered Buoyant Platform
Scott-Lithgow (Offshore) Limited	3.2.4	Tethered Buoyant Platform
Standard Oil Company and Amoco Subsidiaries	3.2.5	Vertically Moored Platform
Continental Oil Company Limited	3.2.6	Tension Leg Platform
Exxon Production Research Company	3.2.7	Submerged Production Platform
Taylor Woodrow Services Limited	3.3.1	Arcolprod
Sir Robert McAlpine and Sons Limited/Consortium	3.3.3	Deep Sea Production System
Sir Robert McAlpine and Sons Limited/Consortium	5a	Plan of Subsea Production Complex
Sir Robert McAlpine and Sons Limited/Consortium	5b	Section of Habitat and Control Module
Westinghouse Electric Corporation	5.1.1	Toroidal Manned Underwater Station

General Dynamics/ Electric Boat Division	5.1.2	Cylindrical Manned Underwater Station
South West Research Institute	5.1.3	Spherical Manned Underwater Station
CanOcean Resources Limited	5.1.5a	Wellhead Cellar
Cameron Iron Works Incorp.	5.1.5b	Wellhead Cellar
Bruker Meerestechnik GmbH	10.3	Mermaid VI Submersible
Perry Oceanographics Inc.	10.3	P.C.16, Submersible
Slingsby Engineering Limited	10.3	L.5, Submersible
United States Navy	10.4.1	Deep Sea Rescue Vehicle
Kockums AB	10.4.2	Underwater Rescue Vehicle
British Oceanics Limited	10.4.3	Taurus
Horton Engineering Limited	10.5.1	Auguste Piccard
Ingenieurkontor Lübeck	10.5.2	Tours 200
D.H.B. Construction Limited	10.6	JIM
D.H.B. Construction Limited	10.6	SAM
O.S.E.L. Group	10.6	WASP

Slingsby Engineering Limited	10.6	SPIDER
Ministry of Defence (N)	11.1.1a	British Submarine Escape and Emmersion Suit
United States Navy	11.1.1b	American Escape and Survival Equipment
Ingenieurkontor Lübeck	11.1.2a	Rescue Sphere
Ingenieurkontor Lübeck	11.1.2b	Location Arrangement
United States Navy	11.2.1	Rescue Procedures: Submersibles
United States Navy	11.2.2	Submarine Rescue Chamber
General Dynamics	12.4	Undersea Power Systems
General Electric	12.4.2	Gemini 1kW Fuel Cell
General Electric	12.4.3a	^{60}Co Organic Rankine System
Westinghouse Electric Corporation	12.4.3b	7.5 kWe Power Plant Module
Rockwell International	12.5.1	Subsea Oilfield Reactor
Rolls Royce and Associates Limited	12.5.2	Seabed Power Generation

University of California	12.5.3a	Critical Mass versus Density
University of California	12.5.3b	Area of Desirable Core Parameters
Sir Robert McAlpine and Sons Limited	17.2.1	Buoyancy Graphs for Containment Vessels

COMPANY	*TABLE NO*	*TITLE*
Dr D.M. Davies, B.P. Limited	6.1.3	External Control
Dr D.M. Davies, B.P. Limited	6.1.4	Internal Control
R. Gartner, Dornier Systems GmbH	7.1.1	Spacecraft Atmosphere Comparisons
Dr D.M. Davies, B.P. Limited	7.2	Effects of Acute Exposure to Carbon Dioxide
Dr D.M. Davies, B.P. Limited	7.4	Major Atmosphere Contaminants and Sources (Submarines)
Dr D.M. Davies, B.P. Limited	7.4.4	Comparative M.P.C.'s for Submarine Contaminants
Ministry of Defence (N)	11.1.1	No Decompression Limits as a Function of Depth (Air)
Sir Robert McAlpine and Sons Limited	12.1	Power Estimates for a Subsea Production Complex

Sir Robert McAlpine and Sons Limited	13.2a	F.A.F.R. Values for Various Occupations
Sir Robert McAlpine and Sons Limited	13.2b	F.A.F.R. Values for Non-Industrial Activities

I should like to acknowledge the financial support provided for this study by the Marine Technology Directorate of the Science and Engineering Research Council as part of the Marine Technology Program at the University of Glasgow.